SPACE, TIME, MATTER, AND FORM

Essays on Aristotle's Physics

OXFORD ARISTOTLE STUDIES

General Editors
Julia Annas and Lindsay Judson

PUBLISHED IN THE SERIES

Space, Time, Matter, and Form
Essays on Aristotle's Physics
David Bostock

Aristotle on Meaning and Essence
David Charles

Time for Aristotle
Ursula Coope

Aristotle on Teleology
Monte Ransome Johnson

On Location
Aristotle's Concept of Place
Benjamin Morison

Order in Multiplicity
Homonymy in the Philosophy of Aristotle
Christopher Shields

Aristotle's Theory of Substance
The Categories *and* Metaphysics *Zeta*
Michael V. Wedin

Aristotle's *De Interpretatione*
Contradiction and Dialectic
C. W. A. Whitaker

Space, Time, Matter, and Form

Essays on Aristotle's Physics

DAVID BOSTOCK

CLARENDON PRESS · OXFORD

OXFORD
UNIVERSITY PRESS

Great Clarendon Street, Oxford OX2 6DP

Oxford University Press is a department of the University of Oxford.
It furthers the University's objective of excellence in research, scholarship,
and education by publishing worldwide in

Oxford New York

Auckland Cape Town Dar es Salaam Hong Kong Karachi
Kuala Lumpur Madrid Melbourne Mexico City Nairobi
New Delhi Shanghai Taipei Toronto

With offices in

Argentina Austria Brazil Chile Czech Republic France Greece
Guatemala Hungary Italy Japan Poland Portugal Singapore
South Korea Switzerland Thailand Turkey Ukraine Vietnam

Published in the United States
by Oxford University Press Inc., New York

First published 2006

British Library Cataloguing in Publication Data
Data available

Library of Congress Cataloging in Publication Data

Bostock, David.
Space, time, matter and form : essays on Aristotle's physics / David Bostock.
p. cm.
Includes bibliographical references and index.
1. Aristotle. Physics. 2. Physics—Early works to 1800. I. Title.
Q151.B67 2006 530.09′01—dc22 2005026313

Typeset by Newgen Imaging Systems (P) Ltd., Chennai, India
Printed in Great Britain
on acid-free paper by
Biddles Ltd., King's Lynn, Norfolk

ISBN 0–19–928686–8 978–0–19–928686–7

1 3 5 7 9 10 8 6 4 2

Preface

Early in my philosophical career, roughly in the time between Geach (1962) and Wiggins (1967), I conceived the ambition of writing a fat tome on Aristotle's conception of substance. As years went by, I worked on it from time to time, and there was quite a long draft in existence before I finally decided to abandon the project. The idea had been that the book would be in three parts: (i) substance as subject of predication (the Logical Works); (ii) substance as what persists through change (the Physical Works); and (iii) substance as what fundamentally exists (the *Metaphysics*). What finally led me to abandon this idea was the realization (not really crystallized until my 1994) that part iii was hopeless. I had begun with the thought that Aristotle's conception of what counted as a substance was interesting and important, even today. To clarify this, I should say that I have never believed in what some philosophers think of as 'metaphysics', i.e. in the idea that philosophers are specially able to describe 'the real nature of the world', as no one else is. That seems to me a mere chimera. But I did believe in what Strawson in the preface to his (1959) called 'descriptive metaphysics', i.e. in the idea that philosophers can at least say how we ordinarily think of 'the world', and what is more fundamental or less fundamental in this ordinary way of thinking. That was where the Aristotelian conception of a substance seemed to me to be highly significant, and yet (in those days) rather neglected.

Of this proposed fat tome, nothing now remains of the work that I had done on part i except what I have recently rescued as my (2004). The essays 1–5 published here represent what is left of the part ii originally planned. They are all concerned, in one way or another, with the idea that substance is what persists through change, taking into account both substance as matter and substance as form. But at the same time the last of them, i.e. essay 5, explains why I did eventually abandon this rather grandiose project. There are *interesting* and *relevant* things to be said about Aristotle's idea of substance, as that appears in the Logical Works and in the Physical Works. But, as I now think, there is almost nothing that is either of contemporary relevance or of antiquarian interest to be found in his *Metaphysics*, and the only conclusion to reach about that work is that confusion reigns. He is perpetually changing his mind, from one page to another, about what 'really' deserves to be called a substance, and the result is that the concept just evaporates. These are hard words, but I mean them: the central books of the *Metaphysics* yield no coherent doctrine at all. I have tried hard to find one (in my 1994), but I have to say that this search did not succeed. Consequently my proposed part iii simply collapsed, and with it all of the original design. The first five essays published or republished here contain what now remains of that original part ii, but each is now presented in a form that is intended to allow it to stand alone, not as part of a continuous discussion.

The remaining essays 6–10 never were part of any overall project, but stem just from my own interest in particular themes that Aristotle treats of in the rest of his *Physics*. (The same interest also explains my contribution to Robin Waterfield's translation of the work (1996), which has something to say about most aspects of the book, but not in any such detail as the essays here.) I cannot pretend that these essays offer anything like a 'complete' coverage of themes from Aristotle's *Physics*, for they concentrate only on selected topics on which I think that I have something of my own to say. In most cases they are simply reprinted here from an earlier publication, though I have (in those cases) added a brief Additional Note to say how I have since changed my mind—or, more often, have not changed it. I hope that it will be found convenient to have them here assembled in one volume.

I thank the relevant publishers for permission to reprint here all of these essays that have been printed before, i.e. essays 1, 2, 3, 7, 9, 10. Full publication details are given in each case as a footnote to the first page of the essay.

New footnotes are indicated by asterisks.

D.B.

Contents

List of Abbreviations

ARISTOTLE'S WORKS

Cael.	*De Caelo*
Cat.	*Categories*
De An.	*De Anima*
De Int.	*De Interpretatione*
EE	*Eudemian Ethics*
EN	*Nicomachean Ethics*
GA	*De Generatione Animalium*
GC	*De Generatione et Corruptione*
Metaph.	*Metaphysics*
Meteor.	*Meteorologica*
PA	*De Partibus Animalium*
Phys.	*Physics*
Post. An.	*Posterior Analytics*

OTHER ABBREVIATIONS

DK	H. Diels, rev. W. Kranz, *Die Fragmente der Vorsokratiker*, 10th edn. (Berlin: Weidmannsche Verlagsbuchhandlung, 1961)
fr.	fragment

1

Aristotle on the Principles of Change in *Physics* I

I

Aristotle opens *Physics* I by stating that an inquiry into nature (*peri phuseōs*), like other inquiries, should begin with an account of the relevant principles (*archai*). He does not tell us what he means by 'nature'—for that we have to wait until book II—and he does not tell us what he means by a 'principle' in this context, but as we read on we may come to think this omission unimportant. For straightway at the beginning of chapter 2 he appears to place himself in the tradition of a series of writers on nature (*peri phuseōs*) whose views on the 'principles' (*archai*) were perfectly well known. Thus Thales held that there was one 'principle', namely water, while Anaximenes selected air and Heraclitus fire; Empedocles again held that there were four principles (earth, water, air, fire), Anaxagoras that there were infinitely many, Leucippus and Democritus that there were just atoms and void, and so on. So Aristotle, it would seem, is preparing to offer us his answer to the question to which these answers had already been propounded by his predecessors: he is preparing to list the ultimate ingredients of the world, and to give an account of how the world is made up from those ingredients. Perhaps this characterization of what the older physicists were up to is rather oversimplified, but I think it is not worth elaborating their problem now. For it soon turns out that Aristotle's problem is after all an entirely different one. The main theme of this paper is to draw attention to the difference, to ask how far Aristotle himself was aware of it, and to trace some of the consequences of his lack of awareness.

We may begin by noticing that as Aristotle's discussion proceeds it soon becomes clear that the principles he is interested in are not so much the principles of natural *objects* (*ta phusei onta*) but rather of natural *processes* or *changes*, and in particular *generations*. This theme enters at the beginning of chapter 4, where the older physicists are said to *generate* things (*gennōsi*, $187^{a}15$) from their

Reprinted from Malcolm Schofield and Martha Craven Nussbaum (eds.), *Language and Logos: Studies in Ancient Greek Philosophy Presented to G. E. L. Owen* (Cambridge: Cambridge University Press, 1982), 179–96.

single body by applying opposites or by separating opposites out of it; and it is firmly established in chapter 5, where the paragraph designed to show that the older physicists were right to rely on opposites (188^{a}30–^{b}26) is precisely an argument that things in general *come to be* from their opposites and *pass away* into them, so the point will hold in particular for the things that *come to be* by nature (*ta phusei gignomena*, ^{b}25). It is the requirements of change and generation, again, which in chapter 6 introduce the idea that some third 'underlying' principle may also be required (189^{a}22–6). And finally, when Aristotle undertakes to develop his own views on the subject in chapter 7, it is a general account of change or generation (*genesis*) that he promises us (189^{b}30–1).

Of course, it is hardly surprising that Aristotle should connect the idea of nature (*phusis*) very closely with that of change, for it has often been pointed out that the older physicists understood nature as a source of change no less than as a principle of static existence,[1] and we know anyway from book II chapter 1 that Aristotle himself understood nature in this way: for him, nature is explicitly a principle of change (*archē kinēseōs*). Besides, he does not entirely lose sight of the original question of the *ingredients* of natural objects. As we shall see more fully later, his account of the 'principles' of change is at the same time intended to reveal the 'principles' of the things that undergo change, fairly much in the sense of the ingredients they are composed of. But it is important to notice that when in chapter 7 Aristotle puts forward his own positive account of change he does not by any means confine himself to *natural* changes or generations. What he promises us is a completely general account of coming to be, and in fact the bulk of his examples are taken from non-natural changes—e.g. a man becoming 'musical' (or better: educated), the generation of a house or a statue, and so on. Clearly the results are intended to apply to natural changes, but his own investigations in chapter 7 are actually of a much wider scope. The inquiry into *nature*, which promised to be a continuation of the speculations in fundamental physics begun by Thales and his successors, has somehow got sidetracked into something altogether more general.

Right from the beginning there has been some hint that the problems of the older physicists will be not so much solved as bypassed, for it is a curious feature of Aristotle's discussion that he often seems much more interested in the question of *how many* principles there are than in the question of *what* they are. The emphasis on the number of the principles is apparent in his opening statement of the problem in chapter 2 (184^{b}15–25), and it can be seen as dictating the strategy of the arguments that follow. For the only thinkers who receive any extended criticism are the Eleatics, who adopted just *one* principle (chapters 2 and 3), and Anaxagoras, who adopted *infinitely many* (chapter 4). Other thinkers are not criticized. Instead Aristotle seeks to extract what is common to them all, and by the time he has finished with his predecessors

[1] See e.g. Mansion (1945: 56–65).

(chapter 5) the only point of this whole discussion that survives is that they all made use of opposites (*enantia*). These opposites were of very different sorts, and sometimes a physicist would make do with just one pair of opposites, and sometimes would invoke several, so the question would certainly seem to arise: *which* pair or pairs of opposites ought really to be adopted as fundamental in our explanation of the physical world? Should we take hot and cold, dense and rare, up and down, love and strife, odd and even, excess and defect, or what? Indeed, this question seems to arise with all the more force in view of the fact that Aristotle apparently claims that only one pair of opposites will be required.

At the beginning of chapter 6 (189^a11–20) he praises Empedocles for having achieved with a limited number what Anaxagoras could do only by using infinitely many opposites, and he goes on to remark that sweet and bitter, white and black, are derivative opposites, which reminds us of the atomistic reduction of these opposites given by Democritus and Plato's *Timaeus*. In the same passage he mentions an argument, which is given more fully at the end of the chapter (189^a13–14, 189^b22–7), and which allegedly establishes that only one pair of opposites will be needed. I will comment on the argument later (pp. 12–13), but clearly the run of the discussion very strongly suggests that Aristotle is wishing to say that in physical enquiry we need take only one opposition as fundamental. This claim is the more surprising when we recall that in his own explanation of the physical world Aristotle employs *two* pairs of opposites to characterize the sublunary elements, viz. hot and cold, wet and dry,[2] and uses a different (triple) opposition when he comes to consider the heavens, viz. motion towards the centre, away from the centre, and round the centre.[3] Even if we suppose (as is perfectly possible[4]) that Aristotle had not yet formulated these theories when he wrote *Physics* I, still it seems outrageous for him to claim, in advance of any empirical enquiry, that only one pair of opposites will be needed. Naturally we shall ask *which* pair of opposites Aristotle here recommends, and it must seem very strange that on the face of it he shows no interest in answering that question.

But this must be to misconstrue his intention. He cannot have meant to put forward the strong claim that only one pair of opposites will be required and then said nothing at all about which they are. So I think we must understand that the particular pair of opposites he has in mind is the pair that emerges as a result of his discussion in chapter 7, namely the pair 'form and privation' (*eidos kai sterēsis*).[5] But *this* alleged pair of opposites is not in any sense a rival to the various pairs employed by the older physicists, and it is quite incongruous to suggest that this is *the* opposition which should be taken as fundamental in physical enquiry. It is as if one were to say that the fundamental opposition in physical enquiry is

[2] See *De Generatione et Corruptione* II. 1–5. The two pairs are explicitly stated to be irreducible at 330^a25–9.

[3] See *De Caelo, passim*, but especially 268^b12–27.

[4] For the early date of *Physics* I see Ross's commentary (Ross 1936: 7).

[5] Note that *Metaphysics I*. 4 is as a whole an argument designed to show that all opposites reduce to ἕξις (or ἕξις τοῦ εἴδους) καὶ στέρησις. See especially 1055^a33 ff.

that between 'a thing and its opposite', for there is no more content to the pair 'form and privation' than this—and in fact there is less, as I shall show later (pp. 9–10). So Aristotle is not after all engaging in physical enquiry himself, as it had seemed from the beginning of the book that he was going to, but rather trying to lay down in advance the general form which any physical enquiry must have. Despite appearances he is not—or should not be—engaging in a dispute with the older physicists as to how many oppositions need to be taken as fundamental in physics, but is rather saying that however many principles the physicist needs to invoke some of them must be classifiable as 'forms' and others as the corresponding 'privations' (and still others as 'underlying things'). Roughly, what is introduced as if it were a continuation of the physicists' investigation of nature has instead become a meta-investigation of the general form which any account of change must take, whether it is an account of *natural* change or not. Commentators have remarked on this change of topic,[6] but I think they have not always noticed that it has some unfortunate consequences for the discussion in chapters 5 and 6. But before I come to this it will be convenient to say something of Aristotle's final doctrine in chapter 7.

As we have just seen, the doctrine is that in any change there will be three 'principles' involved, namely a form, a privation, and an underlying thing (*eidos, sterēsis, hupokeimenon*). This doctrine, properly understood, has the sort of generality which one might expect to result from a purely conceptual investigation, and there is no denying that most of the discussion in chapter 7 seems to be conducted on a conceptual level, indeed one that pays much attention to the niceties of linguistic usage. Thus we begin (189^b32–190^a31) with a detailed account of the kind of change which occurs when a man becomes musical, which is quite unconcerned to discuss the mechanisms and learning processes involved, but wholly devoted to the language we use to describe the change as a whole. Thus we speak of a man becoming musical, of an unmusical thing becoming musical, of an unmusical man becoming a musical man, and so on. Again we speak of a man becoming musical *from* being unmusical, but not *from* being a man. On the other hand we do speak of a statue coming to be *from* bronze, despite the fact that in this case the bronze remains throughout the change, as the man does. Whereas we do not (says Aristotle[7]) speak of the bronze becoming a

[6] e.g. Wieland (1960–1).

[7] 190^a25–6. This is a surprising statement, and I suspect a slip on Aristotle's part. Context demands that we take the Greek ὁ χαλκὸς ἀνδριὰς ἐγένετο in the sense 'the bronze became a statue', but I suspect that Aristotle has been distracted by its other reading 'the bronze statue came to be', and is objecting to this on the ground that the noun-form 'bronze' is inappropriate for use as an adjective, and should be changed to 'bronzen'. (Cf. e.g. *Phys.* 245^b9–246^a4; *Metaph.* 1033^a5–23, 1049^a18–$^b3$.)*

* I now prefer the explanation offered by Code (1976). The text says: 'We say that a statue comes to be from bronze, and not that the bronze becomes a statue.' If we interpret 'not' as 'not only' (as at 190^a5) then sense is restored.

statue. It is pefectly clear throughout these paragraphs that Aristotle is concerned to comment simply on the way we speak, and to show that it conforms to this general scheme:

> When these distinctions are made one can gather from all cases of becoming this point, if one considers them in the way I suggest: namely that there must always be something which underlies and is what becomes, and this thing though numerically one is not the same in form. (I mean 'in form' in the sense of 'in definition': to be a man is not the same as to be unmusical.) The one remains and the other does not; that which is not an opposite remains, for the man remains, but what is not musical (or unmusical) does not remain. (190^{a}13–21)

The terminology is admittedly curious, but the main point seems to be quite clear. Before the change we have an object which can be described as a man (as an underlying thing) or as a thing that is not musical (as having a privation); it is the same thing that is described in these two ways. *Qua* underlying thing it persists throughout the change, in the sense that we have the same man at the end as we had at the beginning, but it can now be described rather as a musical thing (i.e. as having a certain form).

The discussion so far, then, is of a conceptual or linguistic nature, and aims to point out that we use three kinds of concepts—form, privation, and underlying thing—in describing this kind of change. But at the same time we should notice that in the passage just quoted Aristotle claims that the same trio of concepts will apply in *all* cases, and this is a claim which cannot be maintained on the basis of a purely conceptual analysis, as we may see by considering generations *ex nihilo*. The most general form of change, one might say, is simply this: 'At one time it was not the case that *p* and at a later time it was the case that *p*' (and to obtain the form of generation, in our sense of the word, one takes '*p*' as an existential proposition). Now if this is what change is, there is nothing in the concept to rule out generation *ex nihilo*, but it is clear that Aristotle does rule it out. Why? He may of course be influenced by the fact that no physicist had ever seriously embraced this possibility, and since Parmenides all had explicitly denied it (e.g. 187^{a}26–31, 191^{b}13–14), but I do not think he wishes to rest his case on this appeal to authority. Rather, he gives his own argument on the point, and this argument seems to me to rest squarely on empirical investigation.

The crucial passage is 190^{a}31–^{b}10, particularly ^{b}1–10, and runs as follows.

> 190^{a}31 Now coming to be is predicated in many ways. Some things cannot be said to come to be—rather, something is said to come to be *them*[8]—but of substances and of them alone it may be said that they come to be without qualification.

[8] καὶ τῶν μὲν οὐ γίγνεσθαι ἀλλά τόδε τι γίγνεσθαι. To obtain my translation I take τι as subject to γίγνεσθαι and τόδε as complement. (τόδε stands in for, e.g., 'white'.) This sense seems demanded by the next two sentences, though it is not perhaps the most natural way of construing the Greek.

a33 And in other cases it is clear that there must be some underlying thing which is what comes to be. Indeed it is possible to come to be somehow qualified, quantified, related, dated [?], or placed only if something underlies; for only substance is predicated of nothing further, and everything else is predicated of substance.

b1 But if one were to investigate it would become clear that substances too, and all other[9] things which are without qualification, come to be from some underlying thing. There is always something which underlies and from which there comes what comes to be, as for instance plants and animals come from seed.

b5 Things that come to be without qualification come to be either by change of shape, as a statue; or by addition, as those which grow; or by subtraction, as the Hermes from the block of stone; or by composition, as a house; or by alteration, as things which change in matter. But it is evident that whatever comes to be in one of these ways comes to be from some underlying thing.

In this passage Aristotle is arguing that any case of coming to be is a case of coming to be *from* something, so that there is always something that forms the starting point of the change. And the argument which is conducted in the third and fourth paragraphs quoted certainly appears to be an empirical one. This point is perhaps suggested by the opening clause 'if one were to investigate', but it is more strongly indicated by the enumeration in the last paragraph, which is surely not the a priori division of a concept but an empirical collection of cases. The argument hangs, of course, upon the contention that this collection of cases is exhaustive, and it is very difficult to see what a priori grounds one could bring in support of this.

The best way to take this passage, then, would seem to be as an empirical claim that all changes or generations that actually occur are of a certain sort: they are, as we may say, *becomings*, and becoming is distinguished from change in general in that a becoming requires both something which becomes and something which it becomes. Becoming thus includes turning into, growing into, being made into, and so on, but it does not include generation *ex nihilo*, because if that were to occur there would be no underlying thing to function as the starting point of the change. But though it is an empirical claim that all actual changes are becomings, what follows this is I think best viewed as a piece of conceptual analysis. For considering now the mere concept of becoming we may argue that if one thing is properly said to become another then obviously there must be something which does not persist throughout the change, for otherwise there would be no change; but equally there must be something which does persist throughout the change, for otherwise the change would merely consist in one thing coming to be where another had ceased to be, and there would be no reason to say that the one *became* the other. This argument appears to have an a priori certainty, so in *all* cases of becoming we *must* be able to specify something that does persist as well as something that does not.

[9] In view of the first sentence quoted ('and of them alone') commentators generally excise the word 'other', so that the preceding 'and' can be read as 'i.e.'.

I suppose I should admit that this account of two different stages of argument in chapter 7—one empirical and one conceptual—is rather idealized. For one thing, Aristotle very often seems to take no account of the distinction between an empirical and a conceptual enquiry,[10] and certainly he makes no attempt in this passage to draw the distinction as I suggest. For another, he never in fact states the a priori argument I have just supplied him with. But I think it is helpful to recognize that this argument is at work in his mind, for only so can we explain why he is so confident of his conclusion that in any case of becoming there will be something that persists and some form that it acquires or loses. For it is certainly *not* that he has shown us how this doctrine applies in particular cases, nor that its application is entirely straightforward. On the contrary he has discussed only one type of becoming in any detail, namely that typified by a man becoming musical, though he is perfectly well aware that there is another important type of case, namely when a substance comes into being, as when something becomes a tree, or a statue, or vinegar. And where substances are generated it is not always easy to see what it is that remains the same throughout the change. But at this point I should perhaps pause to defend my interpretation of the text, for Charlton[11] has claimed that it is *not* Aristotle's view that when a substance is generated there is always something that persists throughout the becoming.

I must begin by admitting that in my view of chapter 7 there is a serious ambiguity in the phrase 'what underlies' (*to hupokeimenon*), and Charlton's interpretation would avoid this ambiguity. For not until the last chapter of the book—which one may well suspect, for this reason, to be a later addition—do we find Aristotle using his technical term 'matter' (*hulē*) precisely as a technical term for whatever it is that persists, and until then the word seems to bear its ordinary sense of *stuff* or *material.*[12] In my view the expression which Aristotle does here use for what persists is 'what underlies'. But of course that expression is *also* his standard expression for a subject of predication, and Charlton's proposal is to take the expression consistently in the latter sense throughout. In that case what underlies is simply the subject said to become so-and-so, and though this

[10] This emerges in many ways. For one aspect, see Owen (1967) on his use of the phrase *τιθέναι τὰ φαινόμενα.*

[11] Charlton (1970: 77). Charlton's thesis is disputed by Robinson (1974).

[12] Occurrences are at $187^{a}18$–19, $190^{b}9$, 25, and (according to all MSS) $191^{a}10$. The first is in a parenthesis (*καθόλου*, $^{a}16$—*εἴδη*, $^{a}20$) which breaks the line of thought, and it is *possible* that *ὕλη* is there intended technically, but I see no *need* to take it in that way. At $190^{b}9$ *ὕλη cannot* be taken as what persists, for *τὰ τρεπόμενα κατὰ τὴν ὕλην* are obviously things that change in the stuff they are made of, and not—whatever this would mean—things that change 'in respect of what persists'. The point is that their *ὕλη*, i.e. stuff, does *not* persist. At $190^{b}25$ the phrase *ὁ χρυσὸς καὶ ὅλως ἡ ὕλη* is most naturally taken as a way of saying 'gold and any other such stuff', and the same interpretation fits $191^{a}10$. The reason why commentators excise the word in the latter passage is that they think the analogy is designed to explain how the word *ὕλη* is to be understood in its technical sense (for in that case it would be unfortunate to use that same word untechnically in the explanans). But what Aristotle is trying to explain is the phrase *ἡ ὑποκειμένη φύσις* (i.e. *τὸ ὑποκείμενον*), and he has not yet started to use *ὕλη* as a synonym for this.

subject may often persist throughout the change (as when a man is said to become musical) there is no reason to suppose that it always does. Perhaps, then, Aristotle is *only* intending to argue that every change has an underlying thing in the sense of a subject which becomes so-and-so, and is not *also* intending to claim that there is something which persists throughout the change.

One can certainly sympathize with the view that Aristotle's own arguments do not justify a conclusion any stronger than this. One might also sympathize with Charlton's claim (pp. 133–5) that there are several passages elsewhere, especially in the first book of the *De Generatione et Corruptione*, which are (as he puts it) 'not propitious' for interpreting Aristotle as claiming there that there is something which persists when (say) air changes into water or water into earth. But the question at issue is whether Aristotle does, in *Physics* I,[13] claim that there is always something that persists through any change, and it seems to me that the text is quite unambiguous on this point. I have already quoted 190^{a}13–21 (above, p. 5) which says quite unambiguously that in *all* cases of becoming the thing that is not an opposite (e.g. the man) remains. It may perhaps be suggested that Aristotle is writing loosely here; perhaps he means to generalize from the example of the musical man only to other cases in which what is acquired or lost is a quality, quantity, relation, etc., and is still reserving for later treatment the case where a new substance comes into being. After all he has not yet mentioned the generation of substances. But even this defence fails in view of the later passage, 190^{b}9–14. This passage comes immediately after Aristotle has been explicitly discussing the generations of substances, and listing the various ways in which they occur (quoted above, p. 6). He reaches his result—'it is evident that what comes to be in one of these ways comes to be from an underlying thing'—and then at once continues:

> So it is clear from what we have just said that *everything* that comes into being is *always* composite. There is one thing that comes into being, another that comes to be it, and the latter in two ways—either as what underlies or as what is opposite. By the opposite I mean the unmusical thing, and by what underlies I mean the man. And generally the shapelessness, formlessness, and disorder are opposite; the bronze, the stone, and the gold underlie. (190^{b}10–14)

The only ground Aristotle could have for saying that whatever comes into being is composite (*sunthetos*) is that we can distinguish in it two 'elements', one the persisting element (what underlies) and the other the acquired element (the form). If the element said to underlie did not persist in the end product there would be no ground whatever for saying that the end product was composite, and Aristotle explicitly claims that *all* products of becoming are composite. He here makes this claim *immediately after* a passage enumerating the different ways in which substances come into being, and *immediately before* an explanation of

[13] For the reason mentioned earlier (text to n. 12) I shall waive the evidence of *Physics* I. 9, 192^{a}28–34. But clearly in that passage ὕλη is analytically what persists through change.

how the claim applies to substances (i.e. what underlies, and persists in, the statue is its bronze). So he must hold, contrary to Charlton's view, that any substance which comes into being contains both a persisting element and a form. And therefore, as I say, the expression 'what underlies' is being made to do double duty, both for the starting point of the change (i.e. the subject said to become so-and-so) and for what persists throughout the change.

So much, then, in defence of my reading of the doctrine of chapter 7. This doctrine raises a number of important questions, of which the most important seems to be this: is the a priori argument on becoming, which I supplied to support the Aristotelian doctrine, correct? Is it in fact true that in any case of becoming there must be something that persists as well as something that does not? Another question one might well wish to raise, now as a matter of Aristotelian exegesis, is the question whether Aristotle himself consistently espouses this principle, or whether he dropped it as a result of the difficulties that arise in trying to apply it to the case where what comes into being is a substance. But I do not intend to pursue either of these questions here.* Rather, I shall return to the point that Aristotle fails to dissociate his own enquiry from that of the earlier physicists, for this creates some difficulties in chapters 5 and 6.

Aristotle's conclusion is that any becoming can be viewed as a case of one and the same thing persisting all through, but acquiring or losing a certain 'form' (*eidos*); a change is always from form to privation or vice versa. This is *not* to say that all change is, in the traditional sense, between opposites. Equally, the older physicists need not be saddled with the view that *all* change is between opposites, for their practice would rather support the generalization that we shall always have to invoke opposites *when describing the fundamental processes of nature*. Of course, opposites may be fundamental *in physical science* without it following that they are needed to characterize non-natural changes, such as the generation of a house or a statue. What is characteristic of a genuine pair of traditional opposites is that they are opposite ends of a spectrum, an ordering, a scale—e.g. in respect of temperature, density, and so on—and neither is merely the negation of the other. (So you could perhaps represent the traditional view, a bit anachronistically, as the view that in basic physics *quantitative* concepts will be fundamental.) However, form and privation are practically the negations of one another, for anything which is of the right sort to have a certain form but does not have it will be said to have the corresponding privation, and vice versa. Thus form and privation are much more *general* concepts than that of an opposite, but Aristotle seems not to have noticed this point. I say this because in chapter 5 he offers himself to argue *for* the thesis that change is always between opposites (188^{a}31 ff.), though this is not the doctrine of chapter 7. And the argument of chapter 5 is of course mistaken.

* See Additional Note 1.1.

Aristotle begins with the unexceptionable remark that if something becomes white it comes to be white from being *not* white ($188^{a}37$), but then he goes on to add: 'not every case of not white is appropriate here but only cases of black or an intermediate'. One must allow that the Greeks did commonly regard all colours as mixtures of white and black (or better, of pale and dark), but it is obvious that a thing can also come to be white from being colourless, and that is neither opposite to white nor intermediate between white and its opposite. Again if a man ceases to be 'musical' because, for example, he has suffered extensive brain-damage and permanently lost all power of thought, would it be right to say either that he has become 'unmusical' or that he has come to some state intermediate between the two? But the error is clearer still a few lines later, when Aristotle considers the generation of a house or a statue. For with these examples in mind he says ($188^{b}12$–15) that everything that is organized (*hērmosmenon*) must be destroyed by degenerating into disorganization (*eis anarmostian*), and indeed into the *opposite* disorganization. But there is no organization of bricks which is *opposite* to their being organized into a house, and no shape of bronze which is *opposite* to the shape of a statue, because there is no linear ordering of organizations and shapes with that of a house or statue at one end and all others appropriately placed as nearer or further removed from it. Indeed if Aristotle had been thinking clearly he must have seen that this doctrine about opposites is in error, for it is actually incompatible with his own account in chapter 7.

In chapter 7 the concept of form must clearly be taken to cover *any* property which a thing may acquire or lose, with the sole proviso that this acquisition or loss is one that counts as a case of the thing becoming something. It may be suggested that this proviso rules out properties in the category of time, for if something occurs at a certain date, or throughout a certain period, we can hardly represent that as a change in the thing or use the notion of becoming to describe it.[14] But all sorts of other properties will count as forms, and in particular properties *in the category of substance* will count as forms wherever it is possible for them to be acquired or lost (as e.g. when there comes to be a tree, or a house, or a statue, or vinegar). All of these must be counted as forms (or privations), and it is of course standard Aristotelian doctrine that substances *have no opposites* (e.g. *Cat.* $3^{b}24$–32).

The same confusion, between form and privation on the one hand and the traditional pairs of opposites on the other, infects the arguments of chapter 6. In this chapter Aristotle begins to argue that we must recognize a 'third principle' in addition to our opposites, and he is still writing with traditional pairs of opposites in mind. So his first argument is that an opposite (such as density) cannot act on, or make things out of, its opposite (rarity), but must rather act on and make things out of something else which is characterized by that opposite

[14] For this reason the commentators often excise 'or dated' at $190^{a}35$ (quoted above, p. 6). But the slip is quite likely to be Aristotle's.

(189^{a}22–6). The point is here presented in language that is only appropriate to the traditional conception of opposites, for it is surely difficult to conceive of a merely negative privation *acting* on anything, and it seems to be this line of thought which is later illustrated from the thinking of the Milesians and their followers (189^{b}2–8). Nevertheless in this case it would seem reasonable to say that the argument survives the change to the more general concepts of form and privation, for just as Aristotle (wrongly?) thinks of the traditional opposites as predicates, so form and privation too are predicates, and therefore apt to characterize other (underlying) things, and not one another. In fact, by the argument of *Categories* 2^{a}34–^{b}6 every predicate must in the end be predicated of a primary substance, so no inventory of the world's ingredients could be complete if it only mentioned properties. Now it does not automatically follow from this that the subject of our predicates would have to rank as one of the 'principles' we are seeking for, but this is the point which the next argument seems designed to establish.

The main thrust of this next argument (189^{a}27–32) is just the claim that a subject of predication is always prior to its predicates, from which we are invited to infer that it must count as a principle. One thing that is surprising about this argument is that, as Aristotle presents it, it claims that *nothing* that is predicated of a subject can be a principle at all, for the subject would be *the principle of* its predicate, and there cannot be a principle of a principle. If we were right in saying just now that Aristotle's form and privation are both predicative in character, it therefore follows from this argument that they are not principles after all. It seems better, then, not to press the argument to this disagreeably strong conclusion, but to rest content with the claim that a subject of predication must be a principle if its predicate is. (There is, however, no very strong reason to agree with this claim.) A second surprising feature of this argument is that it is introduced by the remark that the opposites are not *the substance of* any existing thing, which presumably must be taken to mean that they are not the substance of anything that Aristotle classes as *a substance*, i.e. that no opposite gives the essential nature (*ti estin*) of any substance. No doubt this may be accepted so far as the traditional opposites are concerned, but we have seen that it does not hold of form and privation, for 'form' must here be taken to include the essential nature (*ti estin*) of any generable substance. As a matter of fact, the point appears to be quite irrelevant to the second argument as stated, but it is crucial to the third.

The third argument (189^{a}32–4) begins by recalling that no substance is opposite to any other substance and then continues 'How then can substance be composed of what is not substance, and how can what is not substance be prior to substance?' The reasoning is somewhat elliptical, but I take it that the first point is that, since no substance is an opposite, if we only admit opposites as principles we shall not have any substance as principle. If so, then the best we could do to explain (the generation of?) substances would be to say that they are somehow made up of opposites that are not substances. But then Aristotle adds that this would be impossible, for what substances are made from would have to be 'prior' to the substances they compose, but nothing else could be 'prior' to substance. Again this argument makes use of the notion of priority in a way that would be hard to justify, but it is clear that the only conclusion to be drawn from it is that our principles must somehow include (a) substance, which they would not do if they consisted just of opposites: the required 'third principle' *must* apparently be (a) substance. But it is not very clear what happens to this argument when we generalize the notion of a pair of opposites to that of a form and its privation. Perhaps if forms may themselves include (secondary) substances, that would satisfy the requirement that the principles must include at least one substance?

The question that thus arises is: does Aristotle in the end wish to endorse the argument of chapter 6 that the missing 'third principle' is *substance*? At first sight it is not clear how much weight we can place on the present paragraph. On the one hand Aristotle does not here commit himself to the cogency of these arguments. He says that they 'provide some support' for the conclusion that there is a third thing (*echein tina logon*, 189^{a}21–2, ^{b}17–18) but adds that the question is still full of difficulty (*aporian echei pollēn*, ^{b}28–9). On the other hand at the end of chapter 7 where he is summing up he refers back to these arguments in a way which apparently commits him to accepting them:

> First we said [sc. in chapter 5] that only the opposites were principles, but then [sc. in chapter 6] that there must also be something else which underlies, and that the principles were therefore three. From what we have said now [sc. in chapter 7] it is clear what sort of opposites they are,[15] how the principles are related to one another, and what the underlying thing is. (191^{a}15–19)

The natural implication of this passage would be that just as chapter 5 established that the principles would at least include opposites, but did not tell us which, so equally chapter 6 established that they would include a third underlying thing, but did not tell us what it was. The difficulty with this line of interpretation, of course, is that the whole drift of the arguments in chapter 6 seems to be that something important has so far been left out, *namely substance*.

Another relevant consideration is that chapter 6 closes with two arguments designed to show that only one pair of opposites will be needed, and Aristotle evidently endorses this conclusion. So you would certainly expect him to endorse the arguments he gives for it, and these arguments both presuppose that there is indeed a 'third principle' while the second of them presupposes further that the 'third principle' is substance. The first of these arguments (189^{b}18–22) is obscure to me, so I here pass over it,* but the second (189^{b}22–7) seems tolerably

[15] τίς ἡ διαφορὰ τῶν ἐναντίων. I take this to mean: what is the differentia distinguishing our opposites from others (so also Charlton 1970: 47). The reference is to the fact that Aristotle's alleged pair of opposites is the (fraudulent) pair 'form and privation', and not, e.g., 'hot and cold'.

* See Additional Note 1.2.

clear. It is claimed that substance is itself a genus, and that there is only one primary opposition within each genus. Clearly this point would be irrelevant unless it was being assumed that the 'third thing', which our opposites are to characterize, is substance.[16] When we add this point to the previous ones, Aristotle's profession to regard the arguments of $189^{a}20–^{b}2$ as tentative does not seem very convincing.

We can develop this line of argument more strongly. It is clear that in chapters 5 and 6 Aristotle represents himself as developing the thought of his predecessors, and in chapter 7 as making a new start and giving us his own views. This would leave it open to us to suggest that in chapters 5 and 6 he is giving us something like a preliminary development of problems (*aporiai*), presenting merely plausible arguments which he does not himself subscribe to. But I think the suggestion is really not very convincing. There can be little doubt that in chapter 5 he is sincerely arguing in his own person for the (mistaken) view that all change is between opposites or an intermediate. So he takes chapter 5 to have established that the principles we are concerned with must include at least one pair of opposites, and the remaining questions are then: how many pairs, and which are they? The answer that he wishes to give to these questions is that the principles include only one pair of opposite—namely the (fraudulent) pair 'form and privation'—and it seems clear that by the end of chapter 6 he takes himself as having established that only one pair of opposites is required. So chapter 5 is not wholly aporematic, and nor is chapter 6, for each concludes with a statement that something has now been established—viz. that the principles must include opposites, and that they include only one pair of opposites—and Aristotle has no wish to go back on these statements.

The arguments concerning the 'third principle' in chapter 6 might be viewed as aporematic on the ground that Aristotle presents them tentatively and ends the chapter by saying that the question is still full of difficulty. But the difficulty is presumably that resolved in chapter 7, $190^{b}23–191^{a}5$—a passage which surely does not reveal any weakness in the arguments of chapter 6. Besides, the arguments that he goes on to give in that chapter for his *other* conclusion, that only one pair of opposites is required, seem to presuppose the correctness of the earlier 'tentative' arguments for the view that there is a third principle, namely substance. One could also note that this conclusion is entirely in harmony with the discussion of chapter 7, where the third principle appears now as *what persists* through change, and all the examples we are offered are substances. At this point the reader may well recall the doctrine of the *Categories* ($4^{a}10–21$) that it is peculiar to substances that they and only they are capable of persisting through change, and he will surely be feeling confident that in Aristotle's own view the

[16] The 'argument' is surely very shaky. In several other places Aristotle states that all oppositions somehow reduce to one fundamental opposition (e.g. *Metaph.* $1004^{a}1–2$, $1055^{a}33$ ff.), but the point is not a very plausible one. The thesis that substance is a genus is denied in *Metaph.* $1053^{b}21–4$, perhaps as a consequence of the many facets of substance revealed in *Metaph. Z–H*.

relevant trio is '*substance*, form, and privation'. I have developed the case for this interpretation at some length because of course our text in fact denies it. Right at the end of chapter 7 we read, to our surprise, 'it is ***not*** yet clear whether the form or the underlying thing is substance' (191^{a}19–20). But surely it is perfectly clear. How could Aristotle in fact avoid the conclusion which he here explicitly disavows?

In fact I see two possible ways of avoiding it while still preserving Aristotle's main doctrine of becoming. One possibility would be to abandon the doctrine of the *Categories* on predication, and to say that there are subjects of (accidental) predication which are not substances; in particular, a form may be predicated of *matter*, and matter is not substance. Evidently *Metaphysics Z*. 3 could be adduced in support of this view. But a more appealing possibility would be to retain the view that forms are always predicated of substances but to deny that that makes substance into the third principle. For according to the most plausible version of the doctrine of chapter 7 the third principle is not what underlies in the sense of what is a subject of predicates (i.e. the thing said to become so-and-so), but rather what underlies in the sense of what persists through change. And this need *not* be a substance (but might, for example, be spatio-temporal continuity, or mass). So here we deliberately split the two senses of 'what underlies', and we abandon a different doctrine of the *Categories*, that only substances persist through change. And we might perhaps cull some support for an Aristotelian version of this alternative by drawing on Aristotle's rather disputed views on 'prime matter'. In fact it is probably fair to say that Aristotle at least toyed with both these lines of thought at one time or another. Of course he also toyed with the thought that chapter 6 was right after all.[17]

Additional Note 1.1

The supposed a priori argument can hardly be based just on the grammar of the verb 'to become' (or its near-synonyms), for I take it that grammar will permit locutions such as '*X*s became extinct in the seventeenth century', or 'As we watched, *X* became smaller and smaller until nothing was left of it at all'. Such claims as these clearly deny that *X* (or *X*s) persisted all through the change, and by themselves they give us no reason to suppose that there was anything else that did persist. So if an a priori argument is to be found we must take it to be based upon claims of the sort '*X* became *Y*', where it is built in that '*X*' refers to something that did exist before the change, and '*Y*' to something that did exist

[17] I am glad to acknowledge my debt to Gwil Owen, whose teaching first led me to an enthusiasm for Greek philosophy and who has encouraged my thoughts on the subject in many ways since then.

after it. Then the idea is that *something* must have persisted all through the change, if it is to count as a case of becoming rather than a case of replacement.

But even if (for the sake of argument) we grant this, still what Aristotle's position requires is something more. For he also insists that the thing which *persists* through the change can in addition be counted as *what underlies* the change, and that would not yet follow. To cite an example given by Charlton (1970: 141), when Midas touches his table, and thereby turns the wood it is made of into gold, one might well say that in this case the thing which persists is *the table*: it existed all through, but was first wooden and then golden. However, we cannot add that the table *underlies* the wood of which it was first made, or the gold that that wood became. Rather, it is the wood that first underlies the table, and then the gold, for it is made of them and not they of it. I change to an example which Aristotle himself did believe in, and which is important to this topic. When (according to him) water changes into earth, what happens is that some matter which is cold and wet changes to matter which is cold and dry, so one thing that remains throughout this change is the coldness. But coldness is one of the opposites, and matter is the subject that underlies it, not vice versa. The doctrine of *Physics* I. 7 is that what remains must underlie both the starting point and the end-point of the change, where 'the underlying nature should be understood by analogy; for as bronze is to a statue, or wood to a bed, or as any shapeless material, before it has acquired shape, is to what does have shape, so too is this underlying nature to the substance, the this, and the being' (191^{a}7–12). It seems obvious that the different 'forms' characterizing Aristotle's four elements—namely the opposites 'hot', 'cold', 'wet', 'dry'—correspond to the *shapes* of such things as beds and statues. So they will not satisfy as things which both persist and underlie a case of elemental change. That is how what is called 'prime' matter becomes relevant. There seems to me to be no prospect of a purely a priori argument that will verify all of what *Physics* I. 7 has to say of change.

So let us revert to a more empirical approach, but continue to simplify by taking into account only those kinds of change that Aristotle here admits. Then there are just two main types to consider, namely 'change of accidents' and 'generation of a substance'. Given Aristotle's overall scheme there is no problem about the first. For an accident must be predicated of a substance, so in this kind of change the same substance persists all through, and changes in its accidental properties. Moreover, it underlies just by being the subject of which these accidental properties are predicated. The problems arise in cases of the second kind, where a new substance comes into being. As we have seen, Aristotle's list of these cases is:

1. by change of shape, as a [bronze] statue;
2. by addition, as those which grow;
3. by subtraction, as a stone statue;
4. by composition, as a house;
5. by alteration, as things which change in matter.

There is no problem with cases 1, 3, and 4, where the persisting things that underlie are fairly clearly the bronze, the lump of stone (or part of it), and the bricks. These underlie in the sense that they are what the statue or the house was originally made *from*, and what they continue to be made *of*, i.e. they underlie in the way that matter does. Perhaps they may *also* be said to underlie in the previous sense, as the subjects of which the new shape or the new arrangement, is predicated, but it is the notion of what a thing is made of or from that is now predominating. The problems are to say what it is that both persists and underlies in cases 2 and 5. Case 2 covers such examples as a seed becoming a tree, and it is not obvious what is supposed to remain the same throughout this change. There is a similar problem with case 5, which covers such examples as wine turning into vinegar, or water turning into air, i.e. where one kind of stuff turns into another.

I deal here in a rather summary way with case 2. To conform to Aristotle's scheme it is simplest to break the change into two distinct phases. In the first some matter, perhaps shaped as an acorn, acquires a new property: it was without life but on germination it comes to acquire life. This change, the acquisition of life, is that which brings a new substance into being. It is to be thought of as a short-lived change, in which we do genuinely have the same matter all through (but perhaps with the addition of a little water).[18] Thereafter there is a second phase, which is a long period of growth, as the living acorn becomes first a seedling, then a sapling, then eventually a mature oak tree. During this growth new matter is absorbed, and old matter is replaced, so that what the thing is made of is changing all the time, and certainly cannot be identified with what it was made from. But what remains the same in this case is just the living thing itself. This is a substance of the traditional kind, and *it* is what grows, changing from being of one size (and shape) to being of another. But it is the same living thing all through, and it underlies by being the persisting subject of which the different sizes are successively predicated.

Case 5 is the subject of the next two essays. In Aristotle's scheme the most basic change from one kind of stuff to another is the transmutation of what he continues to call the four 'elements', i.e. earth, water, air, and fire (though because—in his view—they can and do change into one another, they do not really deserve to be called 'elemental'). What is called 'prime' matter[19] is in the first place the matter that persists through and underlies these transmutations, as it takes on and puts off the four fundamental opposites, hot and cold, wet and dry. So 'prime' matter is what the four elements are made from and remain made

[18] The story is similar, but not quite the same, with the generation of animals. This kind of change is discussed in Essays 4 and 5 below.

[19] The phrase 'prime matter' is mainly due to Aristotle's commentators. He himself does occasionally use 'first matter' in this sense, but more often in the sense of what the commentators call a thing's 'proximate matter', i.e. the specific kind of matter of which the thing is most directly made (e.g. the bronze of a bronze statue). For references see my (1994: 272).

of. But all other kinds of material stuff are constituted by suitable 'mixtures' of more basic stuffs,[20] which in turn are 'mixtures' of yet more basic stuffs, until in the end we come to the elements, which are the most basic kinds of stuff. So all material stuffs are ultimately made of this so-called 'prime' matter. (For if x is made of y, and y is made of z, then x is, indirectly, made of z[21].) Moreover, *everything* in this sublunary world is made of one or more material stuffs, so *everything* is ultimately made of this same 'prime' matter. It appears to follow that prime matter must be capable of taking on any form that anything in this world exemplifies, and hence that it has no form 'of its own'. But is that a comprehensible doctrine? I think that it is, and that Aristotle did himself subscribe to it. Others have thought that it is not, and so have sought to show that it is not what Aristotle believed. That topic is pursued in the next two essays.

But even if Aristotle did believe in prime matter, there is still a question over whether it should be counted as a *substance*. That is perhaps why he concludes his discussion here with 'it is not yet clear whether the form or the underlying thing is substance' (191^{a}19–20). If so, then the 'not yet' might be seen as promising a further discussion of whether matter is substance. However, the next book of the *Physics* opens with a question which is only obliquely related to this, for it asks whether it is more the form or the matter of a natural object that should be counted as its 'nature', and it very soon goes on to equate the nature of a thing with the 'substance' (i.e. essence) of that thing (193^{a}9–^{b}21). For a more direct attack on the question whether matter itself counts as a substance one must look rather to books *Z* and *H* of the *Metaphysics*.

Additional Note 1.2

The argument of 189^{b}18–22 appears to be this. (i) There is no need of more than one underlying thing to be a subject for the opposites. (ii) But if there were two different pairs of opposites, there would also have to be a different underlying thing for each pair. (iii) Unless each of the two [pairs of opposites?[22]]

[20] What Aristotle regards as, properly speaking, a 'mixture' corresponds to what we would regard as a chemical compound rather than a mere mixture. See *GC* I. 10.

[21] Sheldon Cohen (1996) complains that, for Aristotle, being the matter of is not a transitive relation. His example is: 'You can make aluminium out of bauxite, and you can make window frames out of aluminium, but you cannot make window frames out of bauxite—the stuff is too crumbly' (pp. 65–6). The explanation is in this case very simple: aluminium is an element (symbol Al), which is made *from* bauxite, but not made *of* it. (For bauxite is a naturally occurring aluminium oxide, symbol Al_2O_3, and to make aluminium from it you need some process that will drive off the oxygen.) Aristotle does sometimes say that the notion of matter is a relative notion, meaning by this that if x is the *proximate* matter of y, and y the *proximate* matter of z, then x will not be the *proximate* matter of z (e.g. *Metaph. H.* 4, 1044^{a}15–25). But that does not undermine what my text says.

[22] The Greek is ambiguous, and could mean not 'each of the two pairs of opposites' but 'each of the two underlying things'. But this alternative interpretation surely does not yield a more comprehensible argument.

could generate from one another. (iv) But in that case one of the two pairs of opposites would be superfluous.

Suppose, as with Aristotle's eventual theory, we wish to hold that there is just one basic underlying thing, and two basic pairs of opposites, the hot and the cold, the wet and the dry. Then step (iii) *apparently* requires us to admit that one of the four opposites would be able to 'generate from' any of the other three, which would seem to mean that (e.g.) the hot could act not only on the cold but also on the wet and on the dry, thereby making each of them to be hot *rather than* cold or wet or dry respectively. I see no rationale for this claim. But perhaps step (iv) would then follow: since one thing (the hot) has only one opposite (say, the cold), it would then have to follow that wet and dry were intermediates between hot and cold, and could therefore be dropped as redundant.

2

Aristotle on the Transmutation of the Elements in *De Generatione Et Corruptione* I. 1–4

I shall here assume that Aristotle did believe in prime matter, at least for most of the time.[1] In particular he believed that his four elements—earth, water, air, and fire—are each made of the same matter, with different combinations of the four forms 'hot', 'cold', 'wet', and 'dry'. On my account[2] he is committed to this doctrine by his analysis of change in *Physics* I. 7, together with his belief that the four elements can change into one another. I also take it that he affirms the doctrine in *Physics* I. 9 (and elsewhere[3]), when he claims that matter is neither generated nor destroyed, for this is true only of prime matter. I further take it that he affirms the doctrine in *Metaphysics* Z. 3, when he claims that matter has no properties essentially, because again that is true only of prime matter.[4] And finally, I think it is clear that he affirms the doctrine in book II of the *De Generatione et Corruptione* (*GC*), when at the end of the first chapter he sets out his own theory of the elements ($329^a24–^b6$). Several other passages could also be cited from later in that book.[5] On the other side, it is true that for most of *De Caelo* III–IV the argument appears to leave no room for the thesis that the elements share a common matter, and yet even here we find the thesis clearly asserted in a brief passage of IV. 5, namely $312^a30–3$. I think that the best hypothesis here is that this passage is a later addition, reflecting a change of view

Reprinted from C. C. W. Taylor (ed.), *Oxford Studies in Ancient Philosophy*, xiii (Oxford: Oxford University Press, 1995), 217–29.

[1] In modern times doubts over Aristotle's belief in prime matter have been raised by King (1956), Charlton (1970, app.), and Jones (1974). These were answered respectively by Solmsen (1958), Robinson (1974), and Code (1976). Since then the case for doubt has been presented again by Stahl (1981), Furth (1988: 221–7), and Gill (1989, ch. 2 and app.); and the case for orthodoxy by Williams (1982, app.), Cohen (1984), and Graham (1987).

[2] As presented in my (1982), Essay 1 in this volume.

[3] e.g. *Metaph. B.* 4, $999^b12–13$; *H.* 1, $1041^a24–32$; *Λ.* 3, 1069^b35.

[4] One may also *suspect* that Aristotle is thinking mainly of prime matter on some occasions when he says that matter is in itself unknowable (e.g. *Phys.* III. 6, $207^a25–6$; *Metaph. Z.* 10, $1036^a8–9$), or indeterminate (e.g. *Phys.* IV. 2, $209^b9–11$; *Metaph. Z.* 11, 1037^a27; *Θ* 7, 1049^b1).

[5] $332^a17–20$, $35–^b1$; $334^a16–18$, 22–5. See also *Phys.* IV. 9, $217^a20–^b11$.

from the time when those books were first written.[6] But it is not so easy to reach a satisfying view of Aristotle's position in chapters 1–4 of book I of the *GC*. If we assume that he did believe in prime matter when he wrote this discussion, we find some very puzzling passages. Yet if we make the opposite assumption, I think we shall be equally puzzled.

In the first five chapters of the *GC* Aristotle's project is to distinguish and discuss the several varieties of change that he recognizes. Roughly, its plan is that in chapter 1 he criticizes his predecessors, and in chapter 2 his main claim is that generation (*γένεσις*, or more explicitly *γένεσις ἁπλῆ*) should be distinguished from composition or putting together (*σύγκρισις*), contrary to what the Atomists held. Chapter 3 then discusses several problems concerning generation, and chapter 4 returns to the task of distinguishing it from other changes, and in particular from alteration (*ἀλλοίωσις*). Finally, chapter 5 turns to consider growth. The main problem comes with the attempt to distinguish generation from alteration, and it is here that Aristotle seems most clearly to deny the existence of prime matter. This is because he wishes to insist that when one element turns into another we have a genuine case of generation, and not alteration, though it is difficult to see why, if in fact the same matter persists throughout.

It very soon becomes clear that this will be a problem, for the theory of prime matter is of course a monistic theory. But near the beginning of chapter 1 we read:

> Those who say that all things are one, and generate everything from some one thing, must say that generation is alteration, and that what strictly comes to be is altered. (314^{a}8–11)

The pluralists, Aristotle thinks, should take the opposite view, that generation is really a matter of composition, but of the monists he says again:

> Those who construct everything from one must say that generation and destruction are alteration. For according to them the underlying thing remains one and the same, and it is just that kind of thing that we say is altered. (314^{b}2–5)

After some further remarks on the pluralist Empedocles he comes back to repeat the point yet again:

> If there is alteration then the underlying thing must be one element, and there must be one matter of all things that can change into one another; conversely if what underlies is one then there will be alteration. (314^{b}29–315^{a}2)

Since it is clear that Aristotle does *not* think that all generation is really alteration, these remarks apparently commit him to the view that there is *not* one single underlying thing. It is true, of course, that he would not say that his own single

[6] The same hypothesis would also apply to the rather obscure passage a little earlier, at 312^{a}17–21, where again the thesis seems to be asserted.

underlying matter is *an element*, but what difference does that make to the very general argument here propounded?*

Most of chapter 2 is then devoted to a general discussion (and rejection) of atomist doctrines, but at the end Aristotle returns to his main topic and denies that generation is to be regarded as composition. He also gives us, for the first time, a positive characterization of what generation is:

Simple coming to be and ceasing to be do not occur by composition and separation, but when there is change from this to that as a whole. These people [sc. the Atomists] think that all such change is alteration, but there is a difference. For in what underlies there is one thing corresponding to the definition and another corresponding to the matter. When the change is in these, it will be generation or destruction; but when it is in the affections and accidental it will be alteration. (317^a20–7)

Unfortunately, the suggested distinction is ambiguous. The contrast Aristotle means to draw may be between a change in respect of the part of what underlies that corresponds to the definition (*τὸ μὲν κατὰ τὸν λόγον*) and a change in respect of its accidental attributes (*ἐν τοῖς πάθεσι καὶ κατὰ συμβεβηκός*, taking 'and' as 'i.e.'). This is what one might expect him to say from the way that he draws the contrast in *Physics* I. 7, which is by distinguishing between coming to be a substance and coming to have a property in some other category. (In that case he will here be equating the substance of a thing with its definition, as he often does. In fact he repeats just this categorial way of drawing the distinction a little later on in the *GC*, at 319^a11–14.) But on the face of it the present passage suggests a rather different contrast, namely that the underlying thing will be a compound of both matter and form (equating form with definition) and we have a case of generation when *both* of these change, or—as he also puts it—when the thing changes *as a whole*. For it really would be rather strange if, having just distinguished a material and a formal element in the underlying thing, he had said that in generation the change is in *these*, when what he meant was that the change was in just one of them, namely the formal element. But on the other hand we now lose a proper contrast with alteration, for the proper contrast should be that whereas generation changes *both* elements in the thing, alteration changes only one of them. Yet Aristotle obviously does not want to say that every alteration brings about a change either in the definition of the underlying thing or in its matter; rather, he takes it as a change in an attribute that is accidental.

We have, then, two contrasts drawn in this passage: one is a contrast between a change in *both* elements of the underlying thing (the thing as a whole) and a change in only one of them; the other is between a change in its definition (substance) and a change in its accident. Since neither contrast yields an adequate interpretation of the passage on its own, it seems necessary to suppose that Aristotle means both at once. But then we have to bring them into harmony

* See Additional Note 2.1.

with one another, and there is one very natural way of doing this: Aristotle is supposing that when the definition (substance) of an underlying thing changes, then its matter automatically changes too. A change in an accidental attribute will leave the same matter persisting, but a change in definition (in substance) will not: the matter must also become a different matter. That seems to be by far the most natural way of making sense of the passage, but of course it denies the doctrine of prime matter. On this account, when a new substance comes into being, there is *no* matter which persists as the same matter all through.

Chapter 3 offers no further definitions of alteration or generation, but it does raise one issue which is relevant to our topic. Aristotle asks how we are to apply, to the generation of substances, the undeniable principle that what is must in some sense come from what is not. If we take this to imply that what is a substance comes from what is not a substance, we have apparently said that it comes from nothing at all, for where there is no substance there cannot be anything else either ($317^{b}1$–13). He then mentions one of the solutions briefly canvassed in *Physics* I. 8 ($191^{b}27$–9), that what is a substance comes to be from what is potentially, but not actually, a substance, but here he replies that this suggestion does not work. For the supposed thing that is potentially a substance must actually be something, rather than nothing, and that implies that it must actually be a substance after all ($317^{b}14$–36). To help answer this question, he therefore turns to another: if destruction is going on perpetually, why has not the whole world been destroyed long ago? To this the answer is that the destruction of one thing is at the same time the generation of another, and this also answers the original question about what a substance is generated from: it is generated from *another* substance. That is, a substance comes to be from what is not that substance but is a different substance, and the generation of the new substance is at the same time the destruction of the old ($318^{a}1$–18). The relevance of this discussion to our topic is that the *cause* of this cycle is explicitly said to be the underlying matter ($318^{a}9$–11, $319^{a}17$–22); the cycle exists only because this matter can change from one opposite to another. If this is to be relevant to the generation of one element from another, as Aristotle clearly thinks that it is (for it has been one of his main examples), this solution seems to reinstate prime matter after all. Indeed, right at the end of the chapter Aristotle seems explicitly to note this. For he mentions the matter that is the matter of earth and of fire equally, and continues:

And is the matter of each of these different? But if that were so, they would not come into being from one another or from opposites. (For it is fire, earth, water, and air to which the opposites belong.) Or is it that in one way it is the same matter and in another it is different? For the thing itself[7] which underlies is the same, but its being is not the same. ($319^{a}34$–$^{b}4$)

[7] ὅ ποτε ὂν ὑπόκειται, more literally 'that, whatever it is, that underlies'. Aristotle elsewhere uses this idiom to emphasize that he is speaking of the thing which has attributes rather than the thing-*qua*-having the attribute (cf. *PA* $649^{a}15$–19, $^{b}21$–8).

Whether by design or not, this apparently belated recognition of prime matter leads at once to a new account of the difference between generation and alteration, in chapter 4.

Without reference to the way the distinction has already been drawn in chapter 2, Aristotle now accounts for it thus:

Since there is something which underlies, and the affection whose nature is to be predicated of this is something else, and since either of these can change, it is alteration when the underlying thing remains, being perceptible, but changes in its affections (whether they are opposites or intermediate) . . . But when the thing changes as a whole, without anything perceptible remaining as the same underlying thing (for example, when the seed as a whole becomes blood,* or water air, or air water), a case of that sort is generation. (319^{b}8–18)

Here it seems that he has deliberately left a loophole for prime matter: we have a case of generation when nothing *perceptible* remains, but according to him prime matter is not perceptible (332^{a}27, ^{b}1). On the other hand, one has to admit that half a page later, when summing up his doctrine, Aristotle omits this qualification:

When the change of opposites is in quantity, we have increase or decrease; when it is in place, we have locomotion; when it is in an affection and a quality, we have alteration. But when nothing remains of which the other is an affection, or any kind of accident, then we have generation or destruction. (319^{b}32–320^{a}2)

Are we supposed to read in the restriction to *perceptible* remaining things from the fuller statement given just before? Or does Aristotle think that the chapter 2 version, which speaks of the thing changing *as a whole*, does after all leave the same loophole for prime matter? Or should we say that he has once more decided to reject prime matter? Well, certainly, the last alternative seems very improbable, for it looks as though prime matter is affirmed in the very next sentence:

What underlies and receives generation and destruction is matter in the strictest sense, but in a way what underlies the other changes is also matter, because all underlying things receive some kinds of opposite. (320^{a}2–5)

According to this, what is called matter in the strictest sense (ὕλη μάλιστα καὶ κυρίως) is surely regarded as *persisting* through a generation or destruction, for that is the point of saying that it 'receives opposites', as does the matter for any other kind of change. So Aristotle is apparently affirming that some matter does persist through a generation, at the same time as he apparently describes a generation as a change in which nothing persists!

I do not see how Aristotle can be rescued from objection on this point. I guess that what he is thinking is that when, say, earth changes into water, the two

* The theory that the seed becomes blood is not Aristotle's own theory (which I discuss in Essay 5). So one wonders why he gives it here as an example.

termini of this change are just the two substances earth and water. There is admittedly something which remains throughout this change, namely the matter that is common to them both. But earth and water do not count as 'affections' of this matter, nor as 'any kind of accidents' of it (μὴ . . . πάθος ἢ συμβεβηκὸς ὅλως). For they are substances, and so not of a nature to be predicated of any underlying thing.[8] That is a point that one can certainly understand, and might be willing to grant *argumenti causa*, but clearly it does not go far enough. For we can *also* think of the termini of this change as first some matter which is cold and dry, and then some matter which is cold and wet, so that the change is one in which the opposite properties of being dry and being wet are exchanged for one another. But these surely do count as 'affections', or 'some kind of accidents', of the persisting matter; for it is their nature to be predicated of matter, but neither of them could be essential to that matter, if it can 'receive' them both. The objection is that, once a persisting matter is admitted, the change certainly *can* be viewed as an alteration in it, even if it can *also* be viewed in another way as well.

This thought is the beginning of a better conception of our problem. To develop it a little further, let us now observe that it certainly will not do to adopt either of Aristotle's other suggestions, namely that in a generation there is nothing *perceptible* that remains, or nothing that remains the same in *definition*. We can see this at once if we turn our attention from the *GC*, with its focus on fundamental physics, to the more homely and everyday atmosphere of *Physics* I. 7. There generations are appropriately described as changes in which a new substance comes to be, and we are given examples such as: some bronze being moulded into a statue, bricks and timber being put together to form a house, some chips being separated from a block of marble to leave a statue, and so on. These examples are hopelessly at odds with the criteria given in the *GC*. When we make statues or houses we certainly do not use materials which do not remain at all, or which lose their essence during the change, or which are imperceptible. This, it seems, Aristotle has quite forgotten. In his efforts to maintain the thesis that the transmutation of elements is a case of generation, and not a case of alteration (or combination), he has carelessly been led into an account of generation which *only* fits the transmutation of one element into another (or anyway, of one kind of stuff into another kind of stuff).

The fundamental mistake in his discussion in the *GC* is clearly this: he has carelessly taken it for granted that his several varieties of change are mutually exclusive. He has allowed himself to suppose that if a change takes place by alteration, or combination, or separation, then it simply *is not* a generation. His own examples in *Physics* I. 7 show quite clearly that this is a mistake. When some quantity of bronze acquires a new shape, that is obviously a change of accidents,

[8] The claim made at *Metaph.* Z. 3, 1029^{a}23–4, that substances are predicated of their matter, is not Aristotle's considered opinion. This point is argued at length by Lewis (1991, chs. 7, 10). See also my comments on this passage in my translation of, and commentary on, *Metaph. Z* and *H* (Bostock 1994).

so far as the bronze is concerned (in fact it seems to fall under Aristotle's heading 'alteration'). He says himself that houses come into being by combination, and stone statues by separation, and in fact he even says that when one kind of stuff turns into another that is a case of alteration (*ἀλλοιώσει, οἷον τὰ τρεπόμενα κατὰ τὴν ὕλην*, 190^{b}8–9). But he also and at the same time classes them as changes in which a new substance comes into being, i.e. as generations. Here he is surely right. Some changes are *both* changes of accidents *and* generations of new substances.[9] The fundamental error of the *GC* is that (by implication) it denies this. One can see the error in my first quotation from chapter 1:

Those who say that all things are one, and generate everything from some one thing, must say that generation is alteration, and that what strictly comes to be is altered. (314^{a}8–11)

From the thesis that generation is alteration it does not follow that *what* is generated is altered, for the alteration of one thing may be the generation, not of that thing, but of another. So we may diagnose the confusions of *GC* I. 1–4 as arising in this way. Aristotle wishes to maintain that the transmutation of the elements is a case of generation, so the mistake leads him to say that it is not a case of alteration. But that must involve the denial of prime matter, whereas in fact Aristotle has no wish to deny prime matter. It is no wonder that his discussion appears to be incoherent.

When we correct the mistake, we see that the problem has been misconceived, and it should be put in this way: some changes of accidents are *also* generations of new substances, and some are not, but *all* generations of substances must at the same time be changes of another kind. For according to the doctrine of *Physics* I. 7, in any change, including a generation, there is something that was present before the change, remains present through the change, and is still present after it. During the change, then, *this* thing is certainly not generated, but must undergo some other kind of change. The question is whether its change is at the same time the coming into existence of a new substance, that hitherto did not exist at all. But how are we to judge? By what criterion do we decide whether or not a new substance has been brought into being?

Well, the general outline of an answer must surely rely on the familiar connection between substances and criteria of identity, due to Geach, Wiggins, and others. For example, such nouns as 'doctor', 'musician', 'philosopher', and so on behave grammatically just like words for substances, and yet they should not be ranked as predicates in the category of substance. Why not? Well, because where someone is a doctor, we have the same doctor still, or again, if and only if we have the same person, still or again practising medicine. The appropriate criterion of identity in this case is just the criterion of identity for persons, and

[9] This point is very clearly acknowledged at *Phys.* VII. 3, 246^{a}4–9. (But the doctrine of that chapter as a whole is not well thought out.)

doctors do not have their own independent criterion of identity. By contrast, statues and houses do have an identity that is independent of the identity of the material that constitutes them. For example, a house may remain the same house even though a good many bricks are taken away (e.g. to knock two rooms together), others are added, rotting timbers are replaced, the whole thing is reroofed, and so on. Similarly, though perhaps less obviously, with statues. If any visible piece of a bronze statue is replaced, those with an antiquarian interest in the history of art may possibly be unwilling to say that the same statue still remains (though most of us, I think, would not raise objection to a moderate amount of repair and restoration). But there will surely be no objections to invisible repairs. For example, suppose that bronze were a lot less durable than it is, and the bronze of some statue had been eaten away from the *inside* to such an extent that it was in imminent danger of collapse. If, in order to prevent this calamity, a quantity of new bronze were poured into it, without any disturbance to the outer shape and form, then surely all of us would be happy to say that it is still the same statue. But it would not be (all) the same bronze.

There is apparently a much more simple and straightforward way of making this point. The material of which some artefact is made will quite often exist *longer* than the artefact itself: it will already exist when the artefact is created, and will very probably survive the destruction of that artefact. Indeed that very same matter may be made first into one substance and then later into another. If the two have different 'life-histories', they must evidently have different criteria of identity. But this way of putting the point would still allow us to suppose that the artefact is a mere *phase* in the history of its material, so that we have the same house if and only if we have the same bricks, still in the form of a house. If that were all there was to it, then houses would not qualify as independent substances, with their own criterion of identity. So the distinction that we need is the distinction between what is merely a 'phase' of the same continuing substance and what is a new substance, made (initially) of the same continuing matter but with identity-conditions that are independent of it. And now we need to apply this distinction, not to artefacts or living things (as I have just been doing) but to Aristotle's elements, conceived as Aristotle conceived them.

Let us consider a modern analogue. Our own chemical theory is of course very different from Aristotle's in many ways, but still we do admit cases where one and the same substance may exist in different forms, which change into one another in much the same way as Aristotle seems to have thought of his elements as changing. For example, H_2O may exist as water, or as ice, or as a vapour. Moreover, this seems a fair analogue to take, since one supposes that these familiar transformations of H_2O exerted a strong influence on Aristotle's theory. But, from our perspective, the transformations of H_2O are surely *not* generations. That is, one has the same water still, if and only if one has the same H_2O, still in the form of water. Thus the identity-conditions of water are not independent, but simply follow those of H_2O, and water is only a phase in the continued existence of a more basic and underlying substance. To put things another way, when we have some water which is then frozen or vaporized, it ceases to be water, but does not cease to exist. On the contrary, it continues to exist, though not as water, and it may come to exist as water again.

It must be admitted that our ordinary ways of speaking are not quite so unambiguous as I am suggesting. While these remarks about 'water' may seem acceptable enough, that is perhaps partly because we do also tend to use the word as I am supposing that 'H_2O' is used. Thus we say, certainly, that ice is frozen water, but sometimes seem to take this as implying that ice *is* water (or 'is really' water—but no one would say that water 'is really' ice). But if we turn our attention from 'water' to 'ice' the situation seems less clear. Is it so obvious that when ice melts it does not cease to exist? (Certainly, Aristotle claims the opposite at *Metaph. H.* 2, 1042^b27–8.) Or again, suppose that I bought a block of ice from the shop yesterday, took it home, melted it, and then refroze it (say into a different shape). When I am asked, of the new block, 'Is that the ice you bought yesterday?', how should I reply? In order to avoid being misleading, I must of course make it clear that the block I bought yesterday was of a different shape from the one here now, but if I am still pressed to say whether it is, despite that, the same ice, what should I say? It seems that I could say 'Yes—but I melted it and then refroze it', or I could say 'No. I melted that ice down and then froze the resulting water to obtain this'. Either reply seems to me perfectly in order, so far as our ordinary way of talking goes, though only the first is permitted by my claims in the last paragraph. No doubt we would all agree that one has the same ice still if and only if one has the same H_2O, still in the form of ice, but it is not so obvious that one has the same ice again if and only if one has the same H_2O, again in the form of ice.

I admit, then, that in this case ordinary ways of talking are not quite so clear-cut as I have suggested, but nevertheless theoretical considerations seem to me overwhelming. There is no rationale for taking different views of the persistence-conditions of ice and of water, and in other cases where we do not have separate words for the solid and the liquid state of the same substance we have no tendency to suppose that melting the stuff ends its existence. For example, no one would wish to say that the gold of this ring cannot be the same gold as the gold that I bought, just because it has been melted down in the interim. I return, then, to the view as I first stated it. On our theory, H_2O is perhaps a substance, but ice, water, and water-vapour are mere phases of this underlying substance, and have no independent identity-conditions of their own.

Now the way in which Aristotle conceives matter to be related to his four 'simple bodies', earth, water, air, and fire, seems to be exactly parallel to the relation as we see it between H_2O and ice, water, and water-vapour. So it must follow that on his theory earth, water, air, and fire are *not* independent substances, but mere phases of the one basic matter that underlies them all. It genuinely is a substance, but its occurrence in this or that special form does not generate any *new* substance, with its own identity-conditions.

The point applies, I believe, not just to Aristotle's four primary or elemental stuffs, but also to all the other kinds of stuff that he recognizes. Although he conceives of them as formed by the 'mixture' of the four primary stuffs, he certainly does not suppose (as we do) that these mixtures or compounds still contain small particles of the primary stuffs they are composed of. On the contrary, he insists that they are genuinely homoeomerous, and no matter how small a part you take it will contain the same mixture of ingredients as the whole does (*GC* I. 10, 328^a9–16). His suggestion seems to be that just as it is the opposites hot and cold, wet and dry, that characterize the four primary stuffs, so it is these same opposites in different proportions that characterize other stuffs (GC II. 7, 334^b8–31). If this is right, then the other stuffs are essentially the same in composition as the four primary stuffs, and we can conclude by the same reasoning that *none* of them is a substance in its own right. All stuffs are just the basic matter in one or another special form, and the identity of the stuff depends simply on the identity of the matter which is so formed. Thus matter—i.e. what the commentators call prime matter—is the only genuine substance that is a stuff.

But that generalization is not necessary for the main argument, which can be confined just to the transmutation of the four elements earth, water, air, and fire. On Aristotle's general principles, as propounded in *Physics* I. 7, this transmutation requires that all four elements share a common matter, and the transmutation can certainly be viewed as an alteration in that matter. In *GC* I. 1–4 Aristotle denies this, but this is because he has made the mistake of supposing that what can be viewed as an alteration cannot *also* be viewed as a generation, and he has somehow convinced himself that when one element comes from another that *is* a generation. Thus he is led to put forward criteria for being a generation which do not seem to be coherent, and anyway are clearly wrong. When we put this mistake right, and ask instead whether these alterations in matter are *also* generations, the correct answer seems to be that they are not. So here is another mistake on Aristotle's part. But it must be admitted that in order to reach this verdict I have had to rely on there being a close connection between the notion of a substance and the notion of identity over time. Nowadays we see this connection as all-important; it is our only way of understanding what an Aristotelian substance is. But I do not believe that Aristotle himself ever thought in these terms.

Additional Note 2.1

It is obvious that in this discussion of *GC* I. 1 Aristotle is oversimplifying, and the three remarks quoted here cannot be intended quite literally and in full generality, especially the third. (For of course we may regard a tomato that changes from being green to being red as a case of alteration without committing ourselves to the view that there is only one element of which all things are made.)

First, Aristotle must be assuming that a monist will have to take all of space as equally filled with the stuff that is his single element. For example, the Atomists will count as pluralists because they posit not only one matter, of which all atoms are made, but also empty space. However, we can reproduce what is essentially the Atomists' account of change without relying upon empty space. For suppose that there is just one stuff which fills all space (call it 'air'), but which has different properties in different places (say it differs from place to place in 'density' and 'rarity'). Then change can still be explained in the Atomists' way, as always due simply to movement, i.e. movement whereby a piece of the stuff with one property changes place with a piece with the other property.[10] The monist can only be forced to admit that, if there is to be change at all then there must be alteration, if he also accepts that some changes take place in the same place, and without any movement of the stuff involved. But (*a*) this is no longer a point about monism in particular, for the same would apply even to pluralist theories; and (*b*) I remark *ad hominem* that Aristotle's *own* position, as expounded in *Physics* VIII. 7, seems to be that alteration cannot occur without a movement from one place to another (which is what explains why it happened as and when it did, $260^a26–^b15$, $260^b29–261^a13$).

As for the converse claim, that if alteration occurs then monism must be correct, there is surely no way of defending this. Even if all changes whatever are, or are due to, alterations, still no kind of monism will follow. For this we would need the additional premiss that any kind of stuff can change into any other, but then once more it is this extra and unstated premiss that is doing most of the work.

[10] It is interesting to note Descartes's very extreme version of such a theory. On his account there is just one stuff (called 'extension') which fills all space, and the stuff in one place may differ from the stuff in another in one property only, namely their relative motion. (For example, one piece of the stuff may be rotating relative to the stuff surrounding it.)

3

Aristotle's Theory of Matter

I have called this a paper on Aristotle's theory of matter, but it explores Aristotle's own ideas in a way which would have been very surprising to him. The paper is in four sections. In the first I briefly summarize Aristotle's own position, and point to a problem which it involves. I believe that my summary *ought* not to be controversial, though one has to admit that it is nowadays controverted, namely by those who hold that Aristotle never did believe in what is called 'prime' matter. I do not directly engage in argument with these people, but I hope that the reasonableness of the position that I attribute to Aristotle will do something to deflect their scepticism. In the second section of the paper I explore what seems to *us* to be a serious lacuna in Aristotle's own account, namely the problem of providing a suitable 'criterion of identity' for matter. But my conclusion is that, although we can make some headway with this question, still there is no prospect of a full answer along the lines that one might desire. So in my third section I reflect a little on how unsatisfactory this situation really is. Finally, in the fourth section I offer some brief remarks on the eventual demise of Aristotle's theory.

1. ARISTOTLE'S ACCOUNT

A statement of Aristotle's position must begin with his account of change.[1] The first important claim that he makes about this is that there is no such thing as generation *ex nihilo*; on the contrary, in every change there is something to start with, and during the change that thing becomes something which it was not before. It may acquire some accidental property which it previously lacked—for example, it may come to be cold and pale, or thin and under 9 stone, or in Oxford and no longer in London. (These are the kinds of 'change of accidents' that Aristotle standardly recognizes, i.e. change of quality, quantity, or place.[2]) But the more interesting kind of change is one that brings a new substance into being, as when some bronze is moulded into a statue, or when some bricks and

Reprinted from D. Sfendoni-Mentzou, J. Hattiangadi, and D. Johnson (eds.), *Aristotle and Contemporary Science*, ii (New York: Peter Lang, 2001), 3–22.

[1] The *locus classicus* is *Phys.* I. 7. I have discussed this in my (1982), Essay 1 in this volume.

[2] The restriction is argued for in *Phys.* V. 2.

mortar are put together to make a house, or when a seed grows into a plant, or when wine turns to vinegar. But still there is no case, Aristotle insists, in which a change produces something from nothing. There is always something existing beforehand which is what changes, i.e. which comes to be whatever the change results in.

His second important claim is that there is always some way of characterizing what is there at the beginning of the change which allows us to say that that same thing *persists* all through, and is still there at the end.[3] Thus if we start with someone who is pale, and who later becomes tanned by the sun, then (according to Aristotle) we do not say that *the pale one* remains—for certainly that one does not remain pale—but we do say that the same person remains all through this change. Similarly, if we start with a mere jumble of bricks and build a house from them, then the jumble does not remain but the bricks do. *All* changes, Aristotle insists, are like this: there is always something that stays the same throughout the change.

His third claim is: not only is there this something that persists, but also this persistent thing is at the same time a thing that *underlies*. The word is ambiguous. In the Logical Works, to say that *X* underlies *Y* is just to say that *X* is the subject of which *Y* is predicated. In a change of accidents, what persists also underlies in just this sense; e.g. the person is the subject of which the two properties, being pale and being tanned, are successively predicated. But in the Physical Works the phrase also has another sense, for what underlies *Y* may also be identified with what *Y* is 'out of' (ἐκ), which means either what *Y* is made *from* or what *Y* is made *of*. In Aristotle's favourite examples the two meanings coincide, and he does not often distinguish between them.[4] For example, when the bronze statue is made it is made *from* bronze, and as it continues to exist it continues to be made *of* bronze. It is such favourite examples that he leans upon when he tries to explain in general terms how the notion of an underlying thing should be understood:

> The underlying nature should be understood by analogy. For as bronze is to a statue, or wood to a bed—or as any shapeless material, before it acquires shape, is to what does have shape—so too is this underlying nature to the substance, the individual, the thing that is a [basic] existent. (*Physics* I. 7, 191ᵃ7–12)[5]

Aristotle's doctrine, then, is that in any change there is always something which *both* persists *and* underlies, in the sense here indicated.

Now with most of Aristotle's examples of new substances being created one can see easily enough how this doctrine is to be applied. But there are two cases

[3] This interpretation is denied by those who think that Aristotle is not committed to prime matter. (For some references, see e.g. my 1995, n. 1, Essay 2 in this volume.) But it seems to me evidently undeniable.

[4] An exception is *Metaph. H.* 4, 1044ᵃ20–5.

[5] I retain ἡ ὕλη καὶ in *Phys.* 191ᵃ10. The last words translate-or-paraphrase πρὸς οὐσίαν καὶ τὸ τόδε τι καὶ τὸ ὄν.

which he mentions himself in the *Physics*, and which provoke problems. One is the case of a seed growing into a plant, and the other is where one kind of stuff turns into another, as when wine turns into vinegar (*Phys.* 190^b4–5, 8–9). What is it, in these cases, that both persists through the change and underlies what results from the change, as bronze underlies a statue? In this paper I shall not be concerned with the first example, for its importance lies in the implications that it has, not for Aristotle's notion of matter, but for his notion of form. However, the second example, of one stuff turning into another, is crucial to our topic. So I pause here to say something of how Aristotle conceived the notion of a stuff.

First, he does not believe in any kind of atomism, so for him stuffs are homogeneous fillers of space. This is hardly the way we look at things nowadays. We do still speak, of course, of this or that space as being 'filled' by a simple stuff such as water, but yet we also think (*a*) that some sufficiently small parts of that space contain no matter at all, but are just empty, and (*b*) that other small parts do contain matter, but matter that is not water. For nothing less than a whole H_2O molecule can be counted as water.[6] But Aristotle holds that water is strictly 'homoiomerous', which means literally that every part of a body of water is like the whole, and in particular that every part, *however small*, is still water. This has nothing to do with the fact that water is counted by him as an 'element' for he thinks that the same applies to all other stuffs, too.[7] Other stuffs are obtained by mixing elements in various ways, as one might mix salt with water. But such a mixing leads once more to a stuff that is homoiomerous. It does not consist of some parts which are just salt, and some parts which are just water; on the contrary *every* part, however small, is like the whole in being salty water. The 'mixture', if we may still call it that, is in his view completely homogeneous.[8] This, after all, is how things do appear to unaided perception, and Aristotle supposes (naturally enough) that on this point perception is not misleading.

Given this approach to stuffs, how are we to explain the fact that one kind of stuff can turn into another, as when wine turns to vinegar? Aristotle does not suppose that in this transformation anything is added to the wine, or subtracted from it. Rather, if you just leave the wine lying there, without interfering with it in any way, then in time all of what used to be wine turns into all of what is now vinegar. It is, from his point of view, just like the transformation of ice into water from our point of view: you begin with a stuff that has one set of properties, and without adding or subtracting anything you end with a stuff that has another set of properties. But, we said, in any change there must be something which persists, and which also underlies. What can it be? Well, it appears that you have to say that the ice and the water are actually made of *the same stuff*, and this same stuff may exist either in a solid or in a liquid form. Similarly, then, with the wine and the vinegar. Apparently we must say that these, too, are made of *the*

6 I assume here that what 'we' think is what one would find in an elementary school chemistry textbook. I shall persist in this assumption until the last paragraph of the essay.

7 His standard *word* for 'stuffs' is τὰ ὁμοιομερῆ.

8 The *locus classicus* is *GC* I. 10.

same stuff, and it may exist either in the form of wine or in the form of vinegar. And it seems that you must say the same thing *wherever* it turns out that one kind of stuff can change into another.

But this means that, from Aristotle's point of view, you must say this same thing in all cases whatever, for his belief is that any kind of stuff can, in principle, turn into any other. Admittedly there may be need of suitable intermediate metamorphoses. For example, you cannot directly turn blood into stone, or stone into blood. But we can reasonably suppose that stone can be turned into earth, by suitable weathering; that earth can be turned into plants, such as corn, for clearly plants draw their nourishment from the soil; that corn can be turned into bread, as everyone knows; and that in digestion bread is somehow turned into blood. Some, such as Empedocles, would object at this point that plants need water as well as earth, so it is not the earth by itself that turns into corn but only some combination of earth with water and perhaps other things. While Aristotle might accept this as an objection to the detail, he would say that in the long run it makes no difference. For he holds, unlike Empedocles, that earth itself can change into water, and water into earth. So, too, did all the physicists before Empedocles, as well as most who came after.[9] On this point, then, Aristotle adheres to the most long-standing tradition. Though he does indeed speak of the four stuffs, earth, water, air, and fire, as the four 'elements', still he also supposes that they can all change into one another. Their special status as 'elements' appears to be justified by the thought that when one more complex stuff cannot be transformed directly into another, still the transformation will always be possible indirectly, if the complex stuff is first broken down into stuffs that are more elementary, and if necessary into a combination of the four stuffs that are fully elementary, because their composition is the simplest possible.

What is called 'prime' matter,[10] then, is in the first place the stuff of which each of the four elements is made, and in them it is found in its simplest possible forms. The four elements themselves function as matter from which higher stuffs are made, and these in turn are matter for still higher stuffs, and so on. So, in Aristotle's own usage, 'matter' is a relative term, for *X* may be the matter of *Y*, and *Y* the matter of *Z*, while *X* is not counted as the matter of *Z*.[11] But underlying (in this sense) is a transitive relation, so what underlies the four elements does *ultimately* underlie everything else, too. It is with this ultimate underlier (best called not 'prime' but 'ultimate' matter) that I am concerned in this paper.

Aristotle draws several conclusions about it. For example, he argues that there is a sense in which it cannot be generated and cannot be destroyed (*Phys.* I. 9).

[9] An interesting exception is Plato, whose theory in the *Timaeus* allows for the mutual transmutation of water, air, and fire, but requires earth to remain always earth.

[10] The phrase is mainly due to the commentators, rather than to Aristotle himself. But he does occasionally use πρώτη ὕλη in this sense, e.g. *Metaph.* 1015^{a}7–10, 1016^{a}19–20, 1049^{a}24–7 (cf. 1044^{a}25–6).

[11] See e.g. *Metaph. H.* 4, 1044^{a}15–25; *Phys.* II. 2, 194^{b}8–9.

For, by our premisses, if matter were generated it would have to be generated from some pre-existing thing that underlies it. But this, says Aristotle, is 'its own nature', so it would have to be generated from itself, which is absurd. This argument evidently concerns matter as the ultimate underlier. Non-ultimate kinds of matter—wood and water and wax, for example—can certainly be generated, but that is because there is a more ultimate matter that underlies them. But nothing can underlie the ultimate underlier, and so it follows from Aristotle's account of change that it cannot be generated. And for the same reason it cannot be destroyed either.

A second conclusion that he draws is that this ultimate matter has no essence; that is to say, there is no property which it has 'in its own right', and so must continue to have throughout its existence.[12] This was perhaps something of an exaggeration on his part. We can certainly agree, first, that since the same matter can occur in any shape or position or arrangement, no particular shape or position or arrangement can be essential to it. (This applies not just to the ultimate matter, but to all stuffs as Aristotle conceives them.) We can also agree, second, that matter cannot have, as part of its essence, any of the properties which differentiate one specific kind of stuff from another. For example, no piece of matter has to retain the same degree of heat or cold, of hardness or softness, of density and rarity, and so on. For if it can change from being a stuff of one kind to a stuff of another kind, then obviously it must be able to vary in properties such as these. We can add, third, that the same matter need not retain the same volume, or the same weight, or anything of that sort. For example, Aristotle takes it to be an obvious fact of observation that when water turns into 'air' it expands enormously,[13] so clearly the same matter need not always preserve the same volume. If we think in terms of Aristotle's own theory of weight, then the same change will also show that the same matter need not always preserve the same weight. For on his account water has positive weight, which is to say that it has a natural tendency to move downwards until it finds its own level, whereas air has negative weight, which is to say that it has a natural tendency to move upwards until it finds its own level. No amount of the one could possibly be 'the same weight' as any amount of the other.[14]

One can see from these examples why Aristotle should think it plausible to argue that since the same matter is capable of taking on *any* form, there is no form that is essential to it. So it is, as he thinks of it, all things *potentially,* which is just another way of saying that there is nothing which it *has* to be. On some occasions his words do seem to suggest that, since matter is all things potentially, it is *nothing actually*; but this make no sense, and is not his real view.[15] There may perfectly well be a lump of matter that is spherical, i.e. is *actually* spherical; the

[12] The *locus classicus* is *Metaph. Z.* 3. Cf. also *Phys.* III. 6, 207^{a}25–6.

[13] *Cael.* III. 7, 305^{b}11–17.

[14] *Cael.* IV, *passim.*

[15] It is taken to be his real view by some, e.g. by Williams 1982, app. (But Williams agrees with me that it makes no sense.)

point is just that it is not *essentially* spherical, or as Aristotle says it is not spherical 'in its own right'. We may generalize this point. At any time, a given lump of matter must have some definite shape or other, and there is therefore a *determinable* property which it must always retain, namely the property of having *some* shape. Similarly it always retains the determinable properties of having *some* volume, *some* location, *some* degree of temperature, of hardness or softness, and so on. So why should we not count these as part of its essence? Here I guess that Aristotle would reply that he does not recognize such determinable properties as properties at all—he might prefer to think of them as potentialities—or anyway that they are not the kind of properties that can figure in an essence. So a better suggestion might be the property which seems to be the ground of many of those just indicated, namely the property of filling space. Why should that not be counted as part, at least, of the essence of matter? A similar thought would be that though a given piece of matter need not always be moving; still it is always capable of moving, and this is what distinguishes it from the space that it fills.[16] So one might suggest at least the following properties as essential to matter: it fills space completely, i.e. homoiomerously; it is mobile; and it is what all (material) things are ultimately made of. In addition, as I have said, it cannot be created and cannot be destroyed. I shall pay further attention to these ideas in what follows. But Aristotle's official doctrine is that matter has *no* essential properties.

It is no doubt because Aristotle thinks of matter as having no essence that he is doubtful over whether it should be accepted as a substance.[17] If we look at things from a contemporary viewpoint, then the link between substance and essence is indeed quite clear. For simplicity, I state the point here in Geach's terms.[18] Aristotelian substances are paradigmatically things that are identifiable and reidentifiable over time. But any such thing must be of a definite sort, so that there is some sortal noun '*A*', which in some way embodies its criterion of identity over time. The item can then be referred to first as 'this *A*', and later as 'the same *A*', where we do understand what it is for this *A* and that *A* to be (over time) the same *A*. In Aristotelian terms, then, being an *A* would be, or would be part of, the essence of *A*s. So it is the thing's essence that supplies its criterion of identity over time, and thus enables it to be a substance. Hence what has no essence cannot be a substance.

But if we do look at Aristotle's theory of matter in this light, then it must appear to be wholly paradoxical. For we have said that if matter has no essence then it has no criterion of identity over time, and so cannot persist over time, and

[16] *Phys.* I. 9, 192^{a}9–14, alludes to the thought that what Plato in the *Timaeus* calls the receptacle (i.e. space) is assigned the role that Aristotle assigns to matter, in so far as it is what persists and what underlies form.

[17] See *Phys.* I. 7 191^{a}19–20, (cf. 191^{a}12–13, 192^{a}3–6). The interpretation of *Metaph. Z.* 3 on this point is disputed. See e.g. my (1994: 80–1).

[18] Geach (1962, esp. sects. 31–4, 91–2). What I call 'the contemporary viewpoint' originates from Geach's work, but has benefited also from discussions by Wiggins and many others.

so cannot be a substance. Yet matter was introduced to be something that does persist over time, i.e. to be what remains the same when, say, water changes into air. And Aristotle continues to think of matter as persisting over time, even though he denies it an essence. One moral that one can evidently draw is that Aristotle's own perspective is not the same as what I have called the contemporary viewpoint. He apparently sees no connection between essence and identity over time,[19] whereas I would say that this is nowadays a crucial part of our understanding of what an Aristotelian essence is. But the second moral to draw is that, even if Aristotle did not see this as a problem, still it *is* a problem. If matter is to persist over time, then it seems that there must be some appropriate criterion of identity for it. But what is this criterion?

2. TOWARDS A CRITERION OF IDENTITY FOR MATTER

First, I wish to observe that this question still arises even if we drop Aristotle's claim that matter has *no* essence. Certainly, we could admit that the properties I have just listed may be counted as essential properties of matter, but still be puzzled to say what it is for the same matter to be now here and now there. More significantly, we can drop Aristotle's claim that *every* specific kind of stuff can (in principle) change into every other, so that there is only one 'ultimate underlier', and the question will still persist. For suppose (as is, of course, the case) that more careful empirical investigation reveals that there are several equally ultimate kinds of matter, which cannot change into one another, and which may therefore be called 'elements' in roughly our sense.[20] Then presumably each ultimate kind of matter will also have further essential properties, differentiating it from other ultimate kinds; but we may still ask for an appropriate criterion of identity for *each* of these kinds, and it is still not clear how the question is to be answered. Moreover, I think it is clear that if the question can be answered in any one case then effectively the same answer can be given for all other cases, too. I shall return to this point at the end of the present section.

We may usefully begin by introducing the notion of spatio-temporal continuity, which I believe to be a central ingredient in all our understanding of identity over time. I shall say just a little in defence of this claim later on, but meanwhile I merely observe that the idea seems especially plausible in the case of the identity of matter, and I am sure that it would have appealed to Aristotle if it had been brought to his notice. For matter fills space, and while the same matter does not always fill the same space—for it may expand or contract, or move from one position to another—still both we and Aristotle expect all these changes to

[19] It is argued (or assumed) by some that he does share what I call 'the contemporary viewpoint'. I mention, in particular, Frede (1978, 1985).

[20] Aristotle sometimes contemplates this possibility, e.g. at *Metaph. H.* 4, 1044^{a}15–16.

be continuous changes. Let us start, then, with a brief account of how this notion is to be understood.[21]

The essential thought is that it depends upon the continuous movement of boundaries. At the level of ordinary perception stuffs appear as definite lumps, bits, or pieces, separated from their environment by nice sharp boundaries. If stuffs were strictly homoiomerous, as Aristotle supposed, then their boundaries would still be nice and sharp, however microscopically investigated. For simplicity I shall adopt this viewpoint throughout the present discussion. In that case we can say that a *body* is a region of space filled by some stuff, or some combination of stuffs, and marked off from its environment by what I shall call a *physical* boundary, which is a boundary that has the given stuff (or stuffs) on one side of it, and on the other side either some different kind of stuff (or stuffs) or perhaps no stuff at all, but just empty space.[22] A physical boundary, then, is to be distinguished from a merely abstract geometrical boundary, which does not mark any physical distinction, and bodies—as I shall understand the term—are delimited by physical boundaries.[23]

It is now easy to explain spatio-temporal continuity for bodies. The basic idea is that a body x at a time t_1 and a body y at a time t_2 are spatio-temporally continuous if and only if what was the boundary of x moves through space between t_1 and t_2 in a strictly continuous way, until at t_2 it becomes the boundary of y. More precisely, we may put the matter thus: x at t_1 and y at t_2 are spatio-temporally continuous if and only if there is a continuous function f from times to regions of space, defined on all times in the interval $[t_1, t_2]$, taking as values regions which are bounded by a *physical* boundary, with the boundary of x as its value for t_1 and the boundary of y as its value for t_2. To say that the function is continuous is to say that for any specified volume of space, however minute, and for any time t_i, in the interval, there is an earlier time t_{i-n} in the interval (unless $t_i = t_1$) and a later time t_{i+n} in the interval (unless $t_i = t_2$) such that the region associated with t_i differs by less than the specified volume from each region associated with any time t_j between t_i and t_{i-n} or between t_i and t_{i+n}. (Two regions differ by less than a given volume if and only if the sum of the subregions contained in either but excluded from the other has less than that volume.)

In stating this criterion, I have for simplicity assumed that the regions (and hence bodies) that we are concerned with are all bounded by a single surface.

[21] Various objections (which I do not elaborate here) may be raised to definitions proposed by others, e.g. by Strawson (1959: 37), Coburn (1971: 59–60), Swinburne (1981: 19), and Hirsch (1982: 15–21).

[22] Aristotle, of course, denies the possibility of empty space, but that point is of no consequence to the present issue.

[23] One might wish for a more general conception of a *physical* boundary. For example, on Descartes's picture of the universe, the whole of space is filled by exactly the same kind of stuff, but it moves in different ways in different regions. Thus there would be a 'physically significant' boundary between a sphere rotating in one sense, and an enveloping spherical shell rotating in the opposite sense, even though this is not a boundary between different kinds of stuff. The problem will occupy us later.

Evidently it would be simple enough to modify the criterion to allow for bodies bounded by two or more surfaces, as for example a spherical shell has both an inner and an outer surface (and as a lump of Gruyere cheese may be regarded as having one outer surface and many inner surfaces, if we do not want to count the holes as part of the lump). A more interesting variation arises when we allow the regions we are concerned with to be bounded by several outer surfaces, each of which is the boundary of a distinct body. This is certainly relevant when we are considering the notion of 'the same stuff'. For example, the steel which now constitutes one sword may be the same steel as will later constitute another, even though the present sword is subsequently broken into two pieces, from which the new sword is reforged. Following the track of this same steel through space and time, we are not following what is intuitively a *single* spatio-temporal track, but rather what one might call a fission–fusion track, which branches and then comes together again. It is equally straightforward to modify the stated criterion to allow for this, but I shall not delay to state this in detail.

Now spatio-temporal continuity is a special case of what one may call 'causal continuity', whereby the cause or explanation of a thing's present state lies (at least partly) in its previous states, leading up to the present state. At least, it explains the *position* of the body in this way, though not necessarily any of its other states. For the axiom that motion is continuous has held the field for centuries, from Aristotle to Newton and well beyond,[24] and this of course does have the consequence that an object's position at any instant is fully determined by its positions at all previous instants.[25] But there are those who will say that it is really *causal* continuity that matters for identity, not mere spatio-temporal continuity, and that this point applies even in such a relatively simple case as the identity of matter.

To some extent I concede this objection. For example, if we came to believe that motion is in fact discontinuous, proceeding always by very small instantaneous 'jumps', then no doubt all our concepts of identity could survive this change of theory. But also, to some extent I do not concede it, for it seems to me quite basic that, in simple cases at least, identity over time is observable. For example, it is no more difficult for me to tell that this table I am sitting at now is the same table as was here (or nearby) five minutes ago, than it is for me to tell that it is a table. It is true that even in ordinary cases such as this we can find ourselves in difficulties when pushing the concept to see how far it will go, i.e. when trying to say just how much alteration the table can undergo while still

[24] The axiom has had its opponents, as is clear from Aristotle's argument in *Phys.* VI. 1, 231^{b}18–232^{a}18, and the opposition remained for some centuries. (Its course is nicely charted in Sorabji 1983, pt. V.) But in modern times, so far as I am aware, the axiom was not seriously questioned until we come to the theory that electrons 'jump' instantaneously from one 'orbit' to another.

[25] This point relies on Aristotle's own observation that if motion is continuous then there cannot be either first or last instants of motion (*Phys.* VI. 5, 236^{a}13–27). (But I add, against Aristotle, that there can perfectly well be first and last instants of rest.)

remaining the same table, and on this kind of question one cannot but end by agreeing with Hume that there is something 'fictitious' in our notion of identity over time.[26] But in simple cases where, e.g., the table suffers no alteration save change of position, we surely think of identity over time as quite straightforward, because we think of it as observable. And of course the obvious account of what is open to observation is (apparent) spatio-temporal continuity. So this, it seems to me, is actually the root of the concept.

Once the concept is formed it can, no doubt, cut loose from its roots to some extent, and there may be spatio-temporal discontinuities which the causal account will enable us to bridge. Indeed, no living things preserve perfect spatio-temporal continuity, in so far as they all eat and excrete, and may shed bits of themselves in other ways. In some cases, and especially when (as with persons) there is so much else that is important, and that we expect to be conserved, we may well be tempted to abandon spatio-temporal continuity altogether, even in an attenuated form. But this broaches a much more general question than can be discussed here, and I must set it aside to return to our particular question, which is the identity of matter. I set aside, too, thought-experiments on how the concept might be applied in worlds different from ours, for clearly Aristotle's theory of how change occurs must be viewed as a theory about change in this world, not in every possible world. If, then, we retain the assumption that motion is continuous, it is clear that the same matter will in fact follow a spatio-temporally continuous path through space and time, so we have at least made a start on explaining an appropriate criterion of identity for it. But it is only a start. For spatio-temporal continuity is never *by itself* a sufficient criterion of identity for anything, not even for matter.[27]

The crucial point is that spatio-temporal continuity need not be an equivalence-relation. To see this clearly, consider first the simple case of the diverging colour-patches. We can easily imagine beginning with a round orange patch thrown onto a white screen, and then, as we watch, this patch diverges continuously into a round red patch on the one side and a round yellow patch on the other:

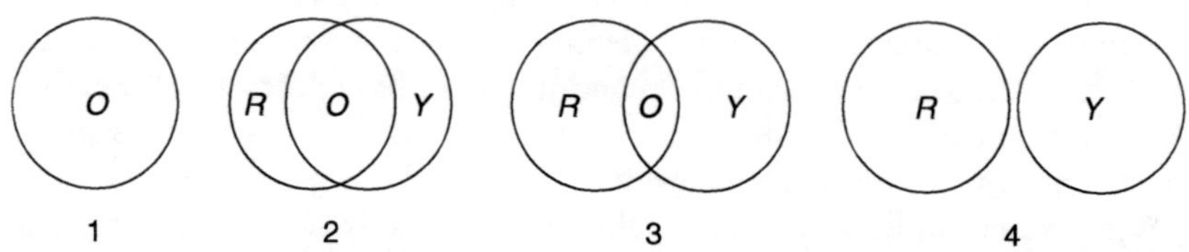

On the account of spatio-temporal continuity that I have given—and I do not see what other kind of account can be given—our original orange patch is in

[26] Hume (1739: I. iv, esp. sect. 6).

[27] There is perhaps one exception, namely atoms (as classically conceived). I say more on this later.

the early stages spatio-temporally continuous with *all* of the following: the remaining orange area, the round area which is partly red and partly orange, the round area which is partly yellow and partly orange, and the whole coloured area. At the end, when the orange area has disappeared, the original orange patch is spatio-temporally continuous both with the resulting red patch and with the resulting yellow patch, and—if we allow fission and fusion—it is also spatio-temporally continuous with the sum of the red and the yellow patches taken together. There would thus be all these different ways of tracing 'the same patch', all equally good so far as mere spatio-temporal continuity is concerned.

Now this illustration concerns two-dimensional light-patches, which are not material bodies, and of course our present concern is with material bodies. But it is easy to give very homely and simple examples of the same phenomenon with material bodies. Suppose that we leave a bowl of water outside on a cold night and in the morning the water is covered by a thick layer of ice. Then (since ice forms from the surface downwards) the water that was in the bowl last night is spatio-temporarily continuous *both* with the water that is still in the bowl this morning *and* with the complex body consisting of the water and the ice together. Mere spatio-temporal continuity cannot tell us which of these resulting bodies consists of 'the same matter' as we began with. In this case we do in fact choose the larger of the two bodies, but it is easy to give a similar case where instead we choose the smaller. For example, suppose that we begin with a puddle of warm oil on a block of ice, and that the oil then begins to melt the ice in contact with it. Then (since oil floats on water) the oil we began with is spatio-temporally continuous *both* with the oil that remains *and* with all the liquid that results, i.e. the oil and the water together. In each of these cases, the spatio-temporal track which we do in fact take to conserve 'the same matter' is one which preserves a body of roughly the same size. But we know that this will not yield the right result in general. For one thing, water does expand when it turns to ice, though only a little. For another, water expands enormously when it turns to 'air', as Aristotle himself knew. Yet it is the same matter which was water and is now 'air'. The general position, then, is this. When we try to follow the 'same matter' using only the criterion of spatio-temporal continuity, it may lead us firmly in one direction for a while; but there are plenty of very ordinary cases where it does not. This is because spatio-temporal continuity, as I have defined it, depends on physical boundaries, and what was one physical boundary may subsequently diverge into two or more, each equally continuous with the original one.

With the examples given one might very reasonably say that we reach the verdict that we do, mainly because we are very familiar with the change from water to ice, and vice versa, and we know how to bring it about, whereas there appears to be no ordinary way of changing oil into water (or into a combination of oil and water). But, one wishes to ask, is there any general principle at work here, which we could apply even to unfamiliar changes? From our contemporary viewpoint, it is very tempting to answer 'yes', for we hold that the spatio-temporal track which preserves always the same matter will also preserve the same *mass*,[28] and that is what determines the issue in our two examples, and in many others that could be given. Thus we have a test for when a specified change does preserve the same matter, and we can in fact apply this test to correct views that we might have been led to by relying on spatio-temporal continuity alone. For example, an apple appears to preserve perfect spatio-temporal contiguity as it first swells upon the tree and then withers on the larder shelf—and that is no doubt an important part of the reason why we regard it as the same apple all through—but the scales soon show us that it does not preserve all the same matter. So this leads us naturally to the following suggestion for a suitable criterion of identity for matter: the same matter is preserved where and only where there is a spatio-temporally continuous track, each cross-section of which is bounded by a physical boundary, and which preserves always the same mass. But there are two reasons why this will not do.

The first and most obvious point to note is that it is only in fairly recent times that we have become convinced of the importance of the notion of mass. Indeed, I think one can fairly say that it was Newton's work which focused attention on this notion, and that it was not until almost a century after Newton—i.e. with the final demise of the 'phlogiston' theory—that it came to be generally accepted that all matter has mass. We cannot just shut our eyes to the fact that for centuries before then people had been prepared to entertain the hypothesis that heat is a form of matter which has no mass (or perhaps a 'negative mass'). Yet they still took themselves to understand what Aristotle meant by 'the same matter'. (Indeed, the alchemists' project of transforming base metals into gold made sense to them just because they were relying on the Aristotelian theory.)

But a second and rather more important point is this: The suggested criterion is not in fact either necessary or sufficient for sameness of matter. It is not necessary, because the same matter need not always occur in distinguishable parcels, marked out from their environment by physical boundaries. Here is a simple illustration. Suppose we have a jug x containing 12 ounces of water, and another jug y containing 10. Suppose then that the contents of these two jugs are amalgamated, and 11 ounces of the resulting mixture is poured off into jug z. Then, relying on our own theory about mass, we will certainly say that some of the water now in jug z was in jug x, and this of course means that some of the water now in jug z is *the same water* as some of the water that was in jug x. But this is a use of identity over time which certainly cannot be explained by the criterion as originally proposed, since there is no relevant *body* of water, first in the one jug and then in the other, to be tested for spatio-temporal continuity. The only relevant fact that we have concerning whole bodies of water is that all of the water in jug z has come from either jug x or jug y. But then, relying just on

[28] I do not here distinguish between inertial and gravitational mass. (The equivalence of the two was noted by Newton.)

the conservation of mass, we can add that not all of it has come from *y*, and this gives us the result. Thus we do make sense of the notion of 'the same matter' even where the criterion suggested has no application.

One's first reaction might be that the point is unimportant. For, so long as the criterion does give the right result where it is applicable, that might seem to be enough. But even this cannot be maintained, as I now show with a different example.[29] Suppose that we have a sealed tray of water, which, however, also has

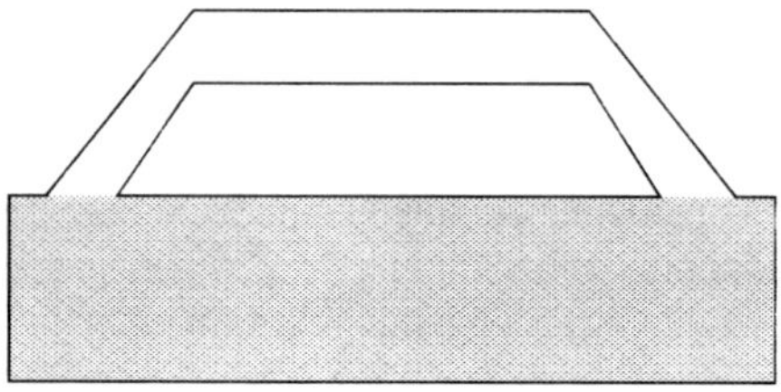

a hollow handle, connecting one end of the tray with the other. Suppose also that, while the tray is filled with water, the handle is filled with water-vapour (i.e. with what Aristotle would regard as 'air'). Both the water in the tray and the vapour in the handle maintain their boundaries in the same place, thus exhibiting spatio-temporal continuity, and each preserves the same mass. So, according to the criterion proposed, each consists of the same matter. But now add that there turns out to be a steady current of the 'air', from one end of the handle to the other, and a smaller but compensating current in the water in the opposite direction.[30] (This motion could be detected, e.g. by the insertion of a small wind-gauge, without disturbing the sealed nature of the apparatus.) It is clear what we should say:[31] at one end of the handle water is turning into 'air' and at the other end 'air' is turning into water, so that the matter is circulating; and neither the water nor the 'air' is made of the same matter throughout.[32] The example should be perfectly comprehensible to the Aristotelian who supposes that water and air are strictly homoiomerous stuffs, and that their boundaries remain sharply defined. It is just that the same matter changes from being water to being air, or vice versa, exactly as it passes through the boundaries that separate water from air. The moral is that even if we are given our modern theory that the same matter preserves always the same mass, still this will not allow us to

[29] The point is noted by Geach in his discussion of Aquinas. See Anscombe and Geach (1963: 69–70).

[30] The explanation might be that the water in the tray is hotter at one end than it is at the other. (But this explanation is irrelevant to the example.)

[31] I mean: the *most probable* hypothesis is clear. Of course there are other *possible* hypotheses, e.g. that air from outside is leaking into the handle at one end and out again at the other, or that new air is being created *ex nihilo* at one end, and destroyed at the other.

[32] Hence, on my account (Bostock 1995, Essay 2 in this volume), the water in the tray is not after all *the same water* throughout. But others may dispute this point.

construct an adequate criterion of identity for 'the same matter', or at least not along the line considered so far. And the reason is essentially the same as before: The same matter, even if it does at one time form a 'body', separated from its environment by a nice physical boundary, need not remain so.

I pause here to recall the point I made at the beginning of this section: The difficulty that we have been encountering has nothing to do with Aristotle's (mistaken) claim that any specific kind of stuff can (in principle) change into any other. For it is clear from my discussion that even if the stuff that we call H_2O[33] were itself an ultimate kind of stuff, capable of just the three forms ice, water, and water-vapour, and unable to change into any other kind of stuff, the same problem over explaining what is to count as 'the same H_2O' would still remain. Indeed, the problem is still there even if *no* specific kind of stuff ever changes into any other. For example, the water within a single glass may be circulating, so that the water in the left half of the glass now is not the same water as the water in that same part of the glass later, but still we have the problem: What is meant here by 'the same water'? The root cause of the difficulty is just the assumption that a stuff (such as water) is strictly homoiomerous, so that one part of it is not separated from another by anything that we can recognize as a *physical* boundary, at least as that notion has been explained so far. Can we, then, offer any alternative explanation?

3. DOING WITHOUT A CRITERION

Reflection on the last examples strongly suggests that what is required is a revision in the notion of a relevant boundary. For the important requirement is not that this is a boundary which separates one specific kind of stuff from another, but that (in addition, or simply instead) it is one which matter does not cross. The same thought occurred to Geach, who says in his discussion of the topic

> We must beware of supposing that 'conservation of mass' is part of what we now *mean* by there being the same stuff in a vessel. It is a matter of common observation that though vessels leak, some leak less than others; the ideal of an unleaky vessel is as easily conceived as that of a rigid body, and it was known, long before there was any idea of 'conservation of mass,' that a hermetically sealed vessel is almost ideally unleaky. 'Conservation of mass' is now known to be a good test for a vessel's 'being unleaky,' or containing the same stuff; but it is not what we *mean* by these expressions . . . [34]

It is clear on only a little reflection that what Geach has in mind by an 'ideally unleaky vessel' is just one that allows no *stuff* to leak in or out. (Thus the vessel may perfectly well be pervious to gravitational fields, or to light or sound or heat,

[33] Of course the description 'H_2O' imports the atomic hypothesis, and this transforms the problem completely, as my final section will observe.

[34] Anscombe and Geach (1963: 69).

so long as these are not themselves reckoned to be forms of matter.) Its role, then, is just to provide a boundary which matter does not cross. But clearly we cannot use this idea to give an *explanation* of what is to count as 'the same matter', for such an explanation would be entirely circular. To say that a boundary is not crossed by matter is just to say that it never does have the same matter now on one side of it and now on the other, and thus it presupposes exactly the notion that it is being used to explain.

I see no way out of this circularity. *If* no specific kind of stuff could change into any other, then a boundary separating one kind of stuff from another would be suitably 'unleaky', but clearly the hypothesis would be quite un-Aristotelian. Again, if we could lay down in advance some independent sign of matter *moving* across a postulated boundary, then that could be used to specify the kind of boundary that we require. From a contemporary perspective this task might not seem impossible, for on our theory any moving matter must have a momentum, and this will generate an effect on any other matter that may be in its way. But, again, this is part of a modern theory, and one can hardly suppose that it would have seemed evident either to Aristotle himself or to his successors over many centuries.

So I think one should look at the situation in this way. Aristotle propounds a theory of matter which consists, basically, in the following claims: (i) Matter fills space, though the same matter may fill now one space and now another. (ii) It is what all material things are ultimately made of. (iii) The same matter may take on various forms (and—on Aristotle's own account—*all* forms). (iv) The same matter is always conserved, which is to say: If we take any boundary which is not crossed by matter over a certain period, then the matter inside that boundary will be the same matter all through the period. Thus, within such a boundary, new matter is never created, or old matter destroyed, either 'out of or into nothing' or by 'increasing or decreasing' the quantity of matter already there. (Of course, matter may increase or decrease in size, or in other respects, but the claim is that this need never be regarded as resulting in more or less matter than there was originally.)

These, I would say, constitute a set of axioms for matter. There are many pertinent issues which they do not determine, most obviously the question of just what is to be counted as made of matter. (Thus, so far as the axioms themselves are concerned, heat and light may or may not be made of matter; that awaits further investigation.) The point that is more relevant to my discussion is that the appropriate criterion of identity for matter is left open. It is constrained, principally by axiom (iv), but not in a way which tells us how to apply it in all cases. Conformably with the general theory, we may entertain various hypotheses on whether a certain boundary is or is not being crossed by matter. What controls these hypotheses is that they must, in the end, allow us to explain the phenomena. (Thus, to revert to my 'sealed tray' experiment, the hypothesis that the 'air' in the handle remains always the same matter seems not to allow us to explain why it exhibits a constant motion in one direction.) In sum, the way in which this general theory is to be applied to the phenomena is not dictated by the theory itself, and the general theory is not falsified by the failure of any one particular attempt to apply it. Nevertheless, it is quite clear that the theory is as a whole an empirical theory, and it will be counted as falsified if there appears to be *no* way of applying it in a particular range of cases.[35]

But how, then, do we attach any meaning to 'the same matter', if this is a notion which the theory uses, but does not explain? The answer is, I think, as Aristotle himself proposed, namely 'by analogy':

> The underlying nature should be understood by analogy. For as bronze is to a statue, or wood to a bed . . . so too is this underlying nature to [the various specific kinds of stuff that are made of it]. (*Phys.* 191^{a}8–13)

Perhaps what Aristotle had in mind in this passage is not exactly the point I am after, which is that we understand the *identity* of matter by analogy with other cases of identity, but certainly it comes close.

As I have urged earlier, there are many cases where we find no difficulty in applying the idea of identity across time, and this applies too to the identity of stuffs. Consider just the examples Aristotle cites here, of how the same bronze may be first a mere lump and later a statue, or how the same wood may be first part of a tree and then a bed. The case of the bronze is very simple, for since we are confident that bronze is not easily transformed into other kinds of stuff, it will be sufficient here if there is a spatio-temporally continuous series of regions, each of which has a boundary that separates what is bronze on one side of it from what is not bronze on the other. The case of the wood is more complicated; since while the wood in question is still part of the tree, its boundaries will not separate what is wood from what is not. But here we can invoke the fact that wood is not internally mobile, so it will be sufficient if the boundaries maintain the same shape and size throughout, and—while they have wood on both sides—also maintain their position relative to the surrounding boundaries that do separate wood from non-wood (i.e. to the boundaries of the whole tree at the beginning, and later to those of the planks into which it is sawn). Of course, it is a *hypothesis* that in each case these boundaries are not crossed by matter, but it is a hypothesis that we have no reason to doubt, and it is surely simpler than any alternative.

In short, there are many cases where we do in practice take it to be quite clear how Aristotle's theory is to be applied, and where it stands up very well to the test of experience. By extrapolation from these cases we understand what the theory demands, even where it is not clear how to apply it. And then, by a kind of 'boot-strapping' move, which I think is actually quite common, consideration of these latter cases gives us an understanding of how we *might* be wrong, even in the cases which were at first accepted as uncontroversial. For it is a very

[35] So I would say that the theory is falsified by the evidence in favour of Einstein's equation '$E = mc^2$'. For what is here represented by 'E' surely cannot be regarded as a kind of matter.

general theory about the nature of the world with which we are concerned, and such theories do not wear on their sleeve how they are to be applied in particular cases.

I am tempted to add that I think that something similar applies to *all* our understanding of identity over time, even where no very general theory is involved—e.g. to the identity of shoes, and ships, and sealing wax, and cabbages, and kings. We *never* have, for things of any kind whatever, what a philosopher will take as a really adequate criterion of identity, i.e. a criterion which can be unambiguously applied even in possible worlds very unlike our own. Even confining attention to this world as it is, our concepts of identity over time show an 'open texture', being centred on some nice clear examples, but with no clear rules on how to extend them to unexpected cases. But that, of course, is too large an issue to be developed here.

4. THE DEMISE OF ARISTOTLE'S THEORY

In Aristotle's own view, the chief rival to his theory of matter was the atomic theory introduced by Democritus (or Leucippus). He very often argues against atomism in all its forms,[36] and while the arguments are hardly convincing from our point of view, still they do have a recognizable appeal to 'common sense'. At any rate, his basic assumption that matter fills space in a strictly homoiomerous way surely is closer to how things appear to us than is the atomist alternative. So too is his claim that motion, space, and time are continuous and not 'atomic'.

Nevertheless, atomism did not die, and eventually it triumphed, though not until Dalton had shown how it really could explain what Aristotle's theory cannot, i.e. why some changes from one kind of stuff to another are possible but others are not.[37] Aristotle would surely regard this as defeat. Yet I would say that it is remarkable how much similarity there is between his ideas and the atomism that succeeded them. I mean to refer here to atoms conceived much as Democritus conceived them, or indeed Newton. Here is Newton's famous description:

> It seems probable to me that God in the Beginning form'd Matter in solid, massy, hard, impenetrable, moveable Particles, of such Sizes and Figures, and with such other Properties, and in such Proportion to Space, as most conduced to the End for which he form'd them; and that these primitive Particles, being Solids, are incomparably harder than any porous Bodies compounded of them; even so very hard, as never to wear or break in pieces; no ordinary Power being able to divide what God Himself made one in the first Creation.[38]

[36] The principal passages are *Phys.* VI (*passim*, but esp. 1–4; 6, 237^a17 ff.; 10); *Cael.* III. 4, 303^a3–$^b3$ (cf. 299^a10–$^b24$); *GC* I. 2. There are many stray references elsewhere (see e.g. Bonitz, *Index*, s.v. ἄτομος), and a curious discussion at *De Sensu* 6, 445^b3–446^a20.

[37] As I observed, you cannot turn stone directly into blood (nor base metals into gold). But why not, on Aristotle's theory?

[38] Newton (1704, book III, query 31; repr. 1952: 400).

Conceived in this way, atoms do fulfil almost exactly the role that Aristotle assigned to matter. Each atom is a homogeneous filler of space separated from its environment by a nice sharp physical boundary, and capable of moving from one place to another. Atoms are neither generated nor destroyed, so that the total quantity of atoms is always conserved. They are what all material things are ultimately made of, and they remain the same through all changes of one kind of stuff into another. Admittedly there is here a major difference from Aristotle's theory, for Aristotle's matter may vary its properties while remaining still the same matter, whereas an atom never varies in its basic properties (e.g. size, shape, mass, or indeed more interesting properties such as valency). A change from one kind of stuff to another must therefore be analysable as some kind of rearrangement of atoms that are in themselves unchangeable, and it is in effect this constraint that allows the atomic theory to explain what Aristotle's theory cannot. (It also gives us a very simple criterion of identity for atoms, for now mere spatio-temporal continuity will by itself suffice, *given* the assumption that the boundary of a single atom never does diverge into two.) So I do not wish to minimize this change of theory, and yet the basic ideas still seem to me to be very Aristotelian. In particular, atoms are made of prime, ultimate, matter—a strictly homoiomerous filler of space. So we may still claim that what ultimately underlies is matter, conceived very much as Aristotle conceived it.

This claim was not overthrown until the present century, when atoms themselves were subjected to decomposition. No doubt it could survive the first steps of that decomposition: when an atom is likened to a miniature solar system, with electrons in orbit around a central nucleus, we can still picture the electrons themselves as 'solid, massy, hard, impenetrable, moveable particles'. But I would say that more modern developments in fundamental physics make it quite clear that this picture does not in fact reflect reality, and we can no longer say that what ultimately underlies is matter. As to what else it is, I can only observe that the physicists are not of one opinion, and I am in no position to have an opinion of my own.

4

Aristotle on Teleology in Nature

Aristotle's main theoretical discussion of teleology in nature is at *Physics* II, chapters 8 and 9, and I take this as my main text. But there are also general and programmatic remarks at the end of *Meteorologica* IV and in *De Partibus Animalium* I, which are both designed as introductions to his Biological Works.[1] And, of course, there is his actual practice in biology, principally in *De Partibus Animalium* II–IV, but some passages in *De Generatione Animalium* are also relevant. The main thrust of the arguments of *Physics* II. 8 appears to be quite clear, so I take this first in Section 2, after some brief preliminary remarks in Section 1. But the overall moral of *Physics* II. 9 is desperately obscure, so before I come to this in Sections 5 and 6 I first beat about the neighbouring bushes. I end in Section 7 with a few general reflections on Aristotle's method of explaining what happens in nature.

1. PRELIMINARIES

By way of introduction I briefly rehearse what has already been said in the *Physics* before II. 8. Book I concerns the analysis of change, whether natural or non-natural, and it claims (in I. 7) that in any change we can distinguish three factors: an underlying thing (ὑποκείμενον) which persists all through the change, what is here called a 'form' (εἶδος) which the underlying thing either acquires or loses during the change, and finally a privation (στέρησις) which is simply the absence of that 'form'. Aristotle ends the discussion by saying—somewhat surprisingly—that while it is clear that there are always these three factors involved in any change, still 'it is not yet clear whether the form or the underlying thing is substance' ($191^{a}19$–20). I think he must mean to be saying (*a*) that this is not yet clear in the case of generations (i.e. changes in which a new substance comes into existence, $190^{a}31$–$^{b}10$), for it seems to be abundantly clear in the other cases he considers; (*b*) that what is not yet clear in these cases is whether form or underlying thing is *the substance of* (= essence of) what is generated, for in these cases 'form' is to be understood in its more usual and much narrower way, i.e. as

[1] I assume without argument that *Meteorologica* IV is a genuine work of Aristotle. If argument is needed, see Düring (1944: 18–27) and Furley (1983).

'substantial form', and what underlies is matter. At any rate, this is the question which *Physics* II takes up in its first chapter.

Book II does begin by limiting attention to natural objects (τὰ φύσει ὄντα) in its opening sentence, and goes on to claim that what is characteristic of them is that they have an internal cause of change or stability (ἀρχὴ κινήσεως καὶ στάσεως, 192^{b}14; ἀρχὴ is soon glossed as αἰτία at ^{b}21). It at once goes on to raise the question whether the 'nature and substance (essence)' of such natural objects (φύσις καὶ οὐσία) is their matter or their form, and proceeds to give arguments on either side (193^{a}9–^{b}18). None of these arguments seem very convincing to us, but I let that pass.[2] At any rate, Aristotle's conclusion (in chapter 2) is that both contribute to the thing's 'nature', and therefore the student of nature must study both (194^{a}12–27). But it is already becoming apparent (*a*) that he thinks form is the more important, and has been unduly neglected by his predecessors (194^{a}18–21), and (*b*) that form is somehow being equated with an end or goal or purpose (τὸ τέλος καὶ τὸ οὗ ἕνεκα, 194^{a}25 ff.). This equation becomes more explicit as we continue.

What follows is the well-known doctrine of 'the four causes' in chapters 3 and 7 (interrupted by a discussion of chance in chapters 4–6). Aristotle claims that there are (i) efficient causes (= what started the change), (ii) final causes (= the goal or purpose of the change), (iii) material causes, and (iv) formal causes. But the discussion is vexing, for whereas efficient and final causes are clearly introduced as causes *of change* it is not clear from what is said why matter and form should be counted as 'causes' at all. At any rate, no suitable 'why'-questions are indicated, to which they might provide the answers.[3] To judge from what we have had earlier, Aristotle would seem to be supposing that they count as 'causes' of the natural changes of a natural object, but none of his examples illustrate this idea. In the case of form, however, we do have a significant claim towards the end of the discussion: 'Three [of the four causes] often coincide; for what a thing is [the formal cause] is the same as what it is for the sake of [the final cause], and that from which the change first began [the efficient cause] is the same as these in form; for man generates man' (198^{a}24–7). My focus in this essay will be on form as final cause, rather than as efficient cause. This brings us to the arguments of chapter 8.

2. THE ARGUMENT OF *PHYSICS* II. 8

The chapter promises to explain why nature is a cause as goal, and then to discuss the role of necessity in natural things. Presumably the promised discussion of

[2] One assumes that the question that Aristotle means to ask is whether it is the thing's matter or its form that is the internal cause of its (natural) changes. It must be admitted that he never quite says this, and *perhaps* it misconstrues his intention. At any rate, the arguments offered on either side in chapter 1 do not bear at all clearly on this notion of an internal cause.

[3] The discussion is vexing in other ways too. There is a succinct account of the main problems in Irwin (1988: 94–102).

necessity is that of chapter 9, but we also hear at once of necessity as the key notion of a *rival* theory that Aristotle sets himself to disprove in chapter 8. According to this theory, when it rains it does so 'of necessity', for what has gone up must be cooled, and what is cooled must become water and descend. Moreover, as this is the full and complete explanation of rain, we can add that it does *not* rain for the sake of anything—neither to swell the corn nor to spoil it (198^{b}16–23).[4] But then the problem is why we should not say the same of the parts of animals, i.e. that some similar necessity makes the front teeth sharp and the back teeth flat, and although it so happens that this is a highly useful arrangement for the animal concerned, still the teeth did not come up this way *because* it is useful, i.e. not for that end (198^{b}23–9). Aristotle says that all his predecessors invoked necessity in this way (198^{b}12), though perhaps we may take Democritus as its chief exponent.[5] But he also refers briefly to Empedocles' theory, that various weird combinations of limbs were at first generated by chance, but those that happened to be useful survived while others perished (198^{b}29–32). This strikes us as a rather different theory, for it reminds us of our own theory of evolution, but apparently Aristotle regards it as just another version of Democritus' theory. This is surely a mistake, for while both theories deny that things in nature happen for a purpose, the one invokes only necessity in its place whereas the other also invokes chance. But perhaps both may be described as materialist theories, in opposition to Aristotle's own, and what he goes on to say is offered as an objection to both. In effect, he puts forward two counter-arguments.

The first is simply that the existing natural arrangements occur always, or anyway for the most part, and that what happens by chance does not. But these natural arrangements certainly appear to be for the sake of something, and since they cannot be set down to chance we must conclude that they are for the sake of something (198^{b}34–199^{a}8).

[4] Whether it is Aristotle's own view that it does not rain for the sake of anything is left unclear. His own first argument (below) would commit him to saying that rain in winter, which is usual, is for the sake of something (whereas rain in summer may not be) only if we are given the extra premiss that rain in winter at least *appears* to be for the sake of something (199^{a}1–2). But the extra premiss seems to me very doubtful. Cf. Irwin (1988: 522). There is a nice discussion of the question in Furley (1985), who cites various opinions on each side. One may suppose that the 'purpose' of rain is either to get water to its natural place in the universe, or to contribute to the constant motion of the sublunary elements, which is their way of imitating the divine. In either case, this 'purpose' applies to all rain, whether in winter or in summer (and what Aristotle here says does not rule this out). The only suggestion that I know of which attributes a 'purpose' to rain in winter, but not to rain in summer, is given by Sedley (1991): he proposes that winter rain has the function of promoting the growth of man's crops, whereas summer rain does not. I cannot believe that Aristotle's view would be so anthropocentric.

[5] For Democritus' reliance on necessity see e.g. *GA* 789^{b}2. It is noted that Empedocles invoked love and hatred as causes, and that Anaxagoras invoked mind as a cause, but Aristotle fairly complains that they did not actually make use of these causes in their detailed explanations (198^{b}14–16; on Anaxagoras cf. Plato, *Phaedo* 97–9).

The second is spelled out at much greater length, and is presented as an analogy between nature and human skill (τέχνη). It will be admitted that, where human skill is concerned, things do come into being for an end or goal, and they come into being in an orderly fashion, one step after another, successively progressing towards that end. But that is just how things come into being by nature too (199^a8–20). Aristotle points to the actions of animals which are just like those of a skilled craftsman, though they should not be credited to any deliberate planning, e.g. the way that a spider builds its web. He goes on to instance the growth of plants—how the roots go downwards for nourishment and the leaves come out above to shelter the fruit. This shows, he claims, that a cause as goal very clearly is to be found in natural processes (199^a20–32). There are, he allows, mistakes in nature, but so also there are in human skill. Occasionally a 'monster' may be born, from some defect in its seed (so he does not utterly rule out Empedocles' weird combinations), but he insists that this is altogether exceptional. To suppose that a seed develops entirely as chance has it is simply to destroy nature altogether; natural objects standardly breed true, and this cannot be by chance (199^a33–$^b26$). Finally he protests that things may perfectly well come about for an end without anything planning this, which he supports by the unexpected claim that skill does not plan either. (He apparently means that the skill itself does not plan, though no doubt its practitioners do.[6] This is not a very helpful point.) Skill and nature, he concludes, work in entirely similar ways, except that in nature the goal is *in* the developing object (199^b26–33).

Let us begin just with the first argument. On the face of it, this is simply an *ignoratio elenchi*. We may admit that there is some absurdity in the view that all the regularities of nature come about merely by chance, but Aristotle's materialist opponents do not wish to say that they do. Perhaps they invoke chance at some points (as Empedocles did), but they *also* invoke necessity, and in their view it is this necessity that accounts for regularity. It may well be that natural regularities do often appear to be goal-directed, but what Aristotle has to show is that this appearance cannot be generated by the working of necessity, in effect by the operation of the laws of matter (as illustrated by the account of the rain). His first argument does not appear to come to grips with the issue at all.

Nevertheless, Cooper (1982, repeated in his 1987) has claimed that the argument is both valid and on target. On his account, Aristotle is relying here on his view that an end or goal is something *good*, and in particular on the thought that it is good for whatever plant or animal is in question to achieve its goal

[6] More commonly the remark is interpreted (after Themistius) as meaning that the skilled craftsman does not need to stop and think what to do next, for he already knows what to do. (Thus Ross, 1936, ad loc., citing *EN* 1112^a34; and e.g. Wieland 1975: 155, Ackrill 1981: 42.) I am inclined to think that this is reading rather a lot into Aristotle's bald statement here, but even if this is his point it remains equally unhelpful. The craftsman does not stop to think because he *knows*, but nature neither thinks nor knows.

(198^b8–9; cf. 195^a23–6).[7] Then the reasoning is supposed to be this. One may grant, at least for the sake of argument, that purely material necessities might account for each of nature's regularities, such as that the horse's teeth always come up in the way that they do, but there would still be something unexplained, namely that this arrangement of teeth is good for the horse. That is, on the materialist's theory it would have to be simply a matter of chance that what the workings of necessity so constantly produced was also good. But then chance has to be invoked for something that happens *always*, and Aristotle is quite right to reject this. If the argument is not convincing to us, Cooper suggests, that is because we do not share Aristotle's view that *always* there have been horses with teeth arranged as now. That is, we do not suppose that the world always has been, and always will be, very much as it is at present.

Now it seems to me unlikely that this was the argument Aristotle intended in his first objection (which is what Cooper is talking of), since his own presentation makes no explicit appeal to the notion of goodness. But, even if it were, the position would not be rectified, since Cooper's argument is quite without force. The trouble is that (like Aristotle) he apparently thinks of the argument as applying individually to each single useful feature, such as the horse's front teeth. The materialist, we are supposing, can give an adequate account of why these teeth are shaped as they are, but he does not explain why this shape is one that is good for the horse. But then, why should not someone else offer a quite different explanation of that (taking into account the horse's way of life), so that nothing is now left to chance? (In a similar way, the meteorologist may hope to explain why it rains a lot in winter, and someone else may try to explain why this is actually a good thing.) In reply to this, Cooper draws on the analysis of Aristotle's account of chance given by Sorabji (1980, ch. 1), and claims that something can still be counted as a coincidence if its explanation can only be given by putting together two quite different explanations that have no connection with one another. (The idea is that there may be one explanation of why I went to market, and another quite independent explanation of why you went to market. Putting these explanations together, we get an explanation of why you and I met. But we still count as having met 'by chance' because the two explanations are independent of one another.) Granting this for the sake of argument, it is then still in some sense a coincidence that the horse's front teeth are shaped in a way that is good for the horse. But to this the materialist's reply is 'Why not?', for coincidences in *this* sense may surely be ones that occur always. For example, in this sense it is a coincidence that wood is useful for building boats and that sand is not. In each case the explanation arises by putting together two quite independent explanations, one accounting for the properties that are needed in suitable boat-building material (and this would stem ultimately from

[7] Gotthelf (1976) is I think alone in arguing that goodness is simply *irrelevant* to Aristotle's final causes.

what boats are wanted for), and the other—no doubt in materialist terms—accounting for the properties that wood and sand actually have. Clearly, coincidences of this kind may be as permanent as you please, and on this reconstruction of the argument the premiss that what happens always cannot be set down to chance must evidently be rejected.

But reflection on this suggested argument of Cooper's reveals another and much better argument. From the materialist's point of view, the laws by which the universe operates pay no attention to goodness and usefulness. They do ensure that there are many regularities in nature, but what one would expect is that some of these regularities should (by 'coincidence') be useful, and some not, in a fairly random manner. However, what is impressively in Aristotle's favour is that *so many* of these regularities evidently are useful for the plant or animal concerned, and how is *that* to be explained on the materialist position? What is difficult to explain is not the isolated occurrence of this or that particular useful feature (such as the shape of the horse's front teeth), which occurs always in the same way, and always usefully. It is rather the fact that practically *all* of the horse's parts are so arranged as to be useful to it. And of course the same applies not just to the horse but to every other kind of animal (and perhaps we may add: to every kind of plant too, though even Aristotle admits that the point is not quite so obvious in the case of plants, $199^{a}23$–30). The point, then, is not that there are *some* cases where nature (always) produces a result that is (always) useful, but that there are *so many*. And surely it is not reasonable to set that down to mere chance?

We, of course, can offer an answer to this problem in terms of the theory of evolution, which is presumably to be regarded as a 'materialist' theory. But Aristotle's opponent has no such theory up his sleeve. No doubt it is fair to remark that Aristotle's background assumptions would anyway have made him unsympathetic to such a theory, even if it had been propounded. For such a theory can hardly be squared with his view that the world has always existed, for an infinity of time past, and does not undergo any grand periodic cycles. But anyway, no such theory had been propounded. As Sorabji points out (1980, ch. 11), there is practically nothing in Empedocles' rather wild speculations to *explain* why animals (and plants) seem to be so well adapted to their environments, and certainly there is no suggestion of the crucial points that (i) animals (and plants) on the whole breed true to type, though (ii) there are minor variations all the time, and perhaps major ones occasionally, 'by chance', and (iii)—an absolutely vital point, without which the theory can explain nothing—not all animals survive to breed, or breed equally prolifically. One can hardly blame Aristotle for not seeing the possibility of a theory along these lines, and the truth is that without it a materialist account of nature *is* bound to be incomplete. Even though it may possibly account for the growth and development of each feature of each animal in terms simply of the laws of matter, still the fact that practically *all* of these features are good and useful for the animal concerned would not be explained. It appears that one must also, in some way, introduce the idea that nature works for an end.[8]

I hesitate to say that this is quite the argument that Aristotle states himself. If anything, it is to be found in his second argument, developing the analogy between nature and human skill. He does there stress the fact that the skilled craftsman will take *many* steps, one after the other and all in the proper order, each making its own contribution to the goal; and nature does the same (199^{a}8–20; cf. 199^{b}28–30). But he does not seem to envisage an opponent who claims that each step taken individually can be fully explained by the laws of matter, its contribution to the overall goal being merely a coincidence. So he rather misses the point that I represent as crucial, namely that while it might possibly be a coincidence that this or that particular step is useful, it can hardly be a coincidence that they *all* are. In other words, what needs an explanation, that the materialist cannot provide, is that practically *all* the parts, of *all* animals, are as if *deliberately designed* to contribute to that animal's good. All the same, it is not difficult to grant that what he does say in this chapter already contains the heart of the 'argument from design'.

In the Christian ages that followed, the argument from design was largely accommodated by supposing that the general laws of matter were not after all neutral with respect to goodness. On the contrary, they had been framed by a benevolent God precisely in order to maximize goodness. To this Aristotle would surely have objected that there is no such *general* notion of goodness as this solution seems to require,[9] and that if the notion was specialized to what was good *for man* (as it generally was) then too much was still left unexplained. For example, we would then have no explanation of why the parts of sharks and crocodiles should be arranged so as to suit the sharks and crocodiles, and not men.[10] To fit Aristotle's views, this type of solution would have to say that the laws of matter are so framed—whether by God, or by some other necessity that we do not understand—that they do work for the good of *every* species of plant and animal. But I believe that Aristotle would have rejected this type of solution outright, on the ground that the characteristics of plants and animals simply cannot be explained by the laws of matter anyway. This is basically because his

[8] Furley (1996) provides an interesting rapprochement between Aristotle's position and the Darwinian theory. This does not assume that species have *evolved*, but does suppose that we need an explanation for their *survival*, and suggests that this is given by the fact that animals are so well adapted to their environments. I wonder whether Aristotle would have been able to see that the survival of a species needs explanation.

[9] Compare *Eudemian Ethics* I. 8; *Nicomachean Ethics* I. 6.

[10] This remark requires two qualifications: (i) There is one place (*Politics* I. 8, 1256^{b}15–22) where Aristotle does suggest that everything in the world exists for the sake of man, but this passage is quite untypical (and quite implausible). (ii) Aristotle does in fact suggest that the shark's mouth is placed underneath partly for the benefit of *its prey* (to give them time to escape), though also for its own benefit (*PA* IV. 13, 696^{b}25–30). But again, this passage is quite untypical. I think it is the only place where he suggests that a feature of one kind of animal is to be explained by the fact that it is good and useful for some *other* kind of animal.

view of the kinds of things that could be explained by the laws of matter was (from our point of view) extremely restricted.

To illustrate this, I leave the *Physics* at this point, for a look at some of the detailed explanations that Aristotle does give in his other writings on nature.

3. ARISTOTLE'S TELEOLOGY IN PRACTICE

Introducing his detailed study of why the various parts of the various kinds of animal are as they are, Aristotle says,

> There are three [kinds of] composition (συνθέσεις). First one may put composition out of what some call the elements, such as earth, air, water, fire. (Or perhaps it is better to call this composition from the powers (δυνάμεις)—not from all of them, but, as has already been stated elsewhere, the wet, the dry, the hot, and the cold are the matter of composite bodies, and the other differentiations follow from these . . .). Second is the composition, from the first things, of the homoiomerous[11] parts of animals, such as bone and flesh and other such parts. Third and last is the composition of the non-homoiomerous parts, such as a face, a hand, and so on. (*De Partibus Animalium* II.1, 646^{a}13–24)

Thus the scheme is that we begin with non-compounded bodies, the four elements (or the four opposites that characterize them, construed here apparently as ingredients of them). From here we progress to certain unspecified compounds, which are very probably inorganic stuffs such as iron or stone or salt or smoke. After this come organic stuffs, possibly first wood and leaf and fruit and grain, and then flesh and blood and bone, i.e. the stuffs of which animal bodies are composed. Out of these in turn are formed the shaped parts of animals, their organs, such as the heart and the liver, or again the face and the hand. And, finally, we have the whole animal.

The formation of both inorganic and organic stuffs is discussed in *Meteorologica* IV, which does not observe any distinction between the two. As Aristotle explains right at the beginning of the book, the hot and the cold are active in this process, whereas the wet and the dry are passive (378^{b}10 ff.), and all his subsequent explanations do indeed invoke heating and cooling. The effect of heat he tells us is 'concoction' (πέψις), which has three main forms: ripening, boiling, and roasting. He admits that these terms have to be somewhat stretched to cover all cases (379^{b}12–17), but clearly cooking is one of his main paradigms. (For example, the digestion of food is a form of cooking similar to boiling (381^{b}3–9), and as a result of this the food is turned into blood.) We need not pause over the details of Aristotle's first tottering steps in the science of chemistry, but simply note his general conclusion to the book, which is at the same time a transition to

[11] This is Aristotle's standard word for a stuff, uniform in composition, i.e. all its parts are like the whole. (Balme 1972 translates as 'homoeomerous', but Lennox 2001*a* prefers 'uniform'.)

the biological works:

Parts of this kind may be generated by heat and cold and the motions due to them, being solidified [sc. from the blood] by the hot and the cold. I mean parts that are homoiomerous, such as flesh, bone, hair, sinew, and the like. For these are all distinguished by the differentiae we have mentioned—for example, stretchability, ductility, fragmentability, hardness, softness, and so on—which are produced by heat and cold and their combined motions. But no one would think that the same applied to the non-homoiomerous parts composed of these, such as a head or a hand or a foot. Similarly cold and heat and their motion will cause bronze or silver to come to be, but do not cause a saw or a cup or a box. In this case skill (τέχνη) is the cause, and in the other case nature or something else. (390^b3–14)

The general position, then, is this. What one might reasonably call the 'laws of matter', from Aristotle's point of view, are primarily concerned with the effects of heating and cooling, and he is here prepared to accept that these laws can account for the production of all different kinds of stuff, both inorganic and organic. But he does not think that they can account for the production of any shaped or structured object, such as the instrumental parts of the body, and to explain these another kind of cause must be invoked. This is evidently the cause which is a thing's form, or its goal (for the two are constantly run together in the biological works).

In fact the standard terminology of *De Partibus Animalium* is to oppose on the one hand 'necessity', which evidently represents the laws of matter, and on the other hand 'nature', which works always for an end. For the most part, Aristotle will invoke *both* of these when explaining some particular bodily feature. For example, man has the hairiest head of all animals.

On the one hand this comes about of necessity, because of the moist nature of his brain and the sutures of the skull—for where there is most moisture and heat, there it is necessary that there should be the largest outgrowth—but on the other hand it is for protection, so that the hair can give shelter from excess of cold or heat (658^b3–8)

There are very many similar passages. Occasionally Aristotle admits that a feature serves no end, and comes about merely of necessity. Thus of the bile in the gall bladder he says,

It seems probable that . . . the bile by the liver is a residue, and not for the sake of anything, like the sediment in the stomach and intestines. Nature sometimes uses even residues to advantage, but still one should not on that account seek a purpose for everything. Rather, when some things are given as such-and-such, many others come about from them of necessity. (677^a13–18)

But for the most part Aristotle gives both kinds of explanation, and clearly does not regard them as in competition with one another.

On the whole and by and large, the way in which the two explanations combine seems to be what one would expect from Aristotle's basic analogy between nature, and skill, and especially from the way in which this was stated at the end of *Meteorologica* IV. The explanation in terms of necessity explains why the right sort of materials are present, and it generally takes the form: when such-and-such matter is already present, in such-and-such conditions, then further matter of a given sort is necessarily produced. The explanation in terms of purpose is then invoked to explain why the material is fashioned into the particular shape and form that it is. This is the implication of the passage on bile just cited, where nature is said often to *use* residues for a good purpose, and the same phraseology often occurs elsewhere. There is an interesting example in the discussion of horns. Here Aristotle first explains what horns are for, and why some animals have horns and others do not (662^{b}23–663^{b}22), and then he goes on: 'We must now say how the necessary nature is (ἡ ἀναγκαία φύσις) in the things that are present of necessity, and used for some purpose by the nature which is in accordance with the definition (ἡ κατὰ τὸν λόγον φύσις)'(663^{b}22–4). Thereupon he tells us of the surplus quantity of earthy matter flowing upwards in an animal, and how nature uses it sometimes for horns, but also sometimes for tusks, or just for better teeth (663^{b}25–664^{a}12). The scheme is, then, that the material is provided by necessity, though in a relatively unstructured state, and it is the end or goal that then determines what precise structure it takes on.

This analysis is broadly in line with that given by Lennox (2001, ch. 8) except that he wishes to add that a further task performed by 'nature' is to ensure that the available material is directed to the appropriate place. So when Aristotle says that the surplus earthy matter 'flows of necessity to the upper region', he wishes us to take *this* occurrence of 'necessity' to refer, not to the necessity that is so often *contrasted* with 'nature as form', but to a necessity that stems from that nature itself. I do not find this interpretation of the present passage very convincing, for it is much more natural to suppose that the upwards flow is here being ascribed to the laws of matter rather than the influence of form. But there are some other passages which suit Lennox's proposal. For example, at 663^{a}29–5 it is said that horn and hoof have the same nature (i.e. the same *material* nature), 'so it is reasonable that where nature has given an excess of this matter to the hooves, in the solid-hoofed animals, it has taken it from above and so made one horn only'. The implication is clearly that nature may direct the same material either upwards for horns (or tusks, etc.) or downwards for stronger hooves, whichever would be the more beneficial. But, in any case, in these passages the material which nature has to work with is provided not by this nature itself, but by what Aristotle calls 'necessity'.

However, the role of necessity is yet further reduced in two passages from *De Generatione Animalium*. In 743^{a}4–^{b}18 Aristotle is discussing the formation, in the embryo, of such stuffs as flesh, sinew, bone, and skin. As we have observed, these are formed from blood by heating and cooling, which have the power 'of necessity' of transforming one kind of stuff into another (743^{a}5–9, 743^{a}36 ff.; there is a clear reference back to *Meteorologica* IV at 743^{a}7). But Aristotle also says that nature *uses* heating and cooling for its purposes, and he concludes, 'all these things should be said to be formed partly of necessity and partly *not* of necessity but for an end' (743^b16–18). The contrast here, then, seems to be that heating and cooling will inevitably have certain effects, but it is due to nature that heating and cooling are *applied* as and when they are, so that the right stuff gets formed in the right place. (Similarly 740^b30–6, and cf. *De Anima* 416^a9–18.) At an earlier passage, 734^b30–6, Aristotle has in fact limited the role of heating and cooling yet further. He there allows that they may well be responsible for such attributes as the hardness and softness of the stuffs formed in this way, but adds that they cannot produce the *logos* by which this is flesh and that is bone. Since the *logos* is here clearly being taken as the *function* of these things, the point would seem to be that mere heating and cooling cannot, for example, make flesh *sentient*, so that it can fulfil its role as the organ of touch.[12] So this too is only to be explained by invoking nature as well as necessity.

To judge from these and similar passages (e.g. *Metaphysics* 984^b11–14; *De Incessu Animalium* 704^b11–15), Aristotle thinks that there is very little about a living thing that can be explained *simply* by invoking the laws of matter. We may add that, given the state of chemistry in his day—or even, one might say, given the state of chemistry at any time before the twentieth century—this judgement is entirely reasonable. Some few facts about the growth and development of living things, or the various activities by which they maintain their life, might be set down to 'the laws of matter', but surely very few. After all, how could mere heat, acting on blood—a *uniform* liquid, according to Aristotle—lead to the generation of an animal embryo of a fixed and definite species? Yet animals are generated out of blood (or, at least, a 'concocted' form of blood, for that is the matter that the female supplies, i.e. the menses). One would surely expect Aristotle to find this inconceivable.

This interpretation of Aristotle's position, i.e. as believing that mere 'laws of matter' *could not* by themselves account for the growth and development of living things, is nowadays quite widely shared.[13] But there are also those who take the opposite view, and think that Aristotle would concede (at least for the sake of argument) that a complete materialist explanation might perhaps be available, and yet still insist that a teleological account was *also* needed.[14] There is certainly some textual support for this position. First, as I have said, Aristotle's actual explanations in *De Partibus Animalium* mostly invoke 'necessity' to explain

[12] This is Aristotle's usual account of the function of flesh. (He does not associate it with muscular action.) But *De Anima* II. 11 modifies the account somewhat.

[13] As proponents one should mention Balme (1972: 76–84, 93–101; 1987: 281), Gotthelf (1976), Waterlow (1982, ch. 2, esp. 86–92), Cooper (1982, 1985), Bradie and Miller (1984), Lear (1988, ch. 2, sect. 3), Furth (1988, sects. 10–12), and Cohen (1996, ch. 5, sect. 1).

[14] The chief advocates are Nussbaum (1978, interpretive essay 1), Sorabji (1980, chs. 9–11), and Charles (1991); one might add Boylan (1981), Lewis (1988), Freudenthal (1995, ch. 1). Irwin (1988, ch. 5) thinks that this would be the best course for Aristotle to take, but does not claim that he took it.

one thing (the availability of suitable matter) and 'nature' to explain *another* (why that matter acquires the particular shape and structure that it does). This division of labour is explicit in many places (as in the passage on horns, cited earlier), and is often easily imposed even where it is not made explicit (as in the passage on human hair). But there are a few exceptions, where Aristotle seems to credit exactly the same effect both to 'necessity' and to 'nature's purposes'. For example, deer are unusual both in having solid horns and in shedding their horns 'of necessity, owing to the weight of the horns, and for the sake of advantage, in order to be lightened' (*PA* 663^{b}13–14). The explanation in terms of necessity is here quite unconvincing, but still Aristotle does give it, and does not seem to see it as conflicting with the other explanation in terms of advantage. It seems clear that he does not rule out the possibility of exactly the same thing being fully explained by two quite different 'causes', i.e. both by a material necessity and at the same time by nature's purposes.[15]

It is *possible* that he means to insist upon it, especially in what he says in his later biological writings of the so-called connate *pneuma* (σύμφυτον πνεῦμα). This is apparently a material substance, of a warm and gaseous nature, which is found in the blood in the form of frothy bubbles.[16] It is assigned a crucial role in many of the things that are associated with 'soul' or 'form'. For example, it transmits sensations from the sense-organs to the heart, and it transmits motion from the heart to the limbs. It also plays a central part in Aristotle's account of the generation of animals, for it is this that transfers form from the male parent to the matter supplied by the female, which it does by conveying certain 'movements' from the one to the other. So far, these functions of the connate *pneuma* are compatible with the view that it acts merely as an instrument, which is *used* by the soul (i.e. form) to accomplish its various purposes, in ways that

[15] Of all the many explanations given in *De Partibus Animalium* and *De Generatione Animalium*, I count only six which seem clearly to say that exactly the same feature is due both to necessity and to nature's purposes, namely: *PA* 663^{b}12–14 (on why deer shed their horns, quoted in the text), 679^{a}25–30 (on why the squid squirts ink), *GA* 739^{b}27–31 (on why the embryonic sac is formed), 755^{a}17–30 (on why fish eggs swell), 788^{b}30–789^{a}3 (on why the front teeth are formed before the back teeth), 789^{a}9–14 (on why first teeth are shed). But there are a good many others that *could* be taken in this way. (For a different kind of example, recall *Phys*. II. 7, 198^{a}19–20: why did they go to war? Because they had been raided (efficient cause), and in order to gain control (final cause). The text does not suggest that one cause explains one feature of the effect and the other explains another.)

[16] For general discussions of the 'connate *pneuma*' see e.g. Peck (1942, app. B), Balme (1972: 158–65), Nussbaum (1978, essay 3), Freudenthal (1995, ch. 3). One expects a body (σῶμα) that is warm and gaseous to be a suitable compound of Aristotle's elements fire and air, but in *GA* II. 3, 736^{b}30 ff., we are told that this *pneuma* is 'more divine' than the familiar four elements, and is 'analogous' to the fifth element which is the matter of the heavenly bodies. The point of the analogy is apparently that the heat of this *pneuma* can give rise to animals, and so can the heat of the sun (i.e. in spontaneous generation), whereas ordinary fire does not. From this Aristotle infers that its heat 'is not fire and does not originate from fire' (737^{a}1–8). These remarks certainly *seem* to imply that Aristotle is now revising his view that all the matter of the sublunary world arises from a suitable combination of the four familiar elements. (The implication is denied by Balme 1972: 160–4, but accepted by Charles 1988, and Freudenthal 1995, esp. 34–5.)

could not be explained by any simple material necessity. But this *pneuma* has another role too, in Aristotle's theory of spontaneous generation, in which it seems that it cannot be regarded as merely the 'instrument' of form.

Aristotle believes that certain animals are generated 'spontaneously', i.e. without the influence of any parent of the same species.[17] He thinks that there are species of shellfish, e.g. oysters, which are *only* generated in this way. His account of how it happens (given mainly in *GA* III. 11) is roughly this: the heat of the sun acts on suitable 'parcels' of earth and water mixed, to generate *pneuma* in them. The heat in this *pneuma* then somehow generates life in the parcel of matter, and it develops into an animal of a familiar kind. Which kind of animal it becomes seems to be settled simply by the ratio of the mixture of earth and water that we began with. (Perhaps it also makes a difference just how much heat the sun provides.) It is utterly mysterious (to me) how Aristotle could have thought this mechanism would give rise always to animals of a familiar species, and not just to an Empedoclean mixture of all kinds of living parts, probably with no suitability to one another. For there is in these cases nothing to provide an indwelling form which might guide and structure the development of the animal. But evidently he did think it. So he must have supposed that in spontaneous generation it is simply 'the laws of matter'—including the laws of *pneuma*, for that apparently is a kind of matter—that explain *all* features of the resulting animal. And if that applies to oysters and whelks and sea-urchins, then why should it not apply to all other animals too? (For Aristotle evidently does not think that these shellfish lack any internal structure, or that their structure fails to fit them for their environment and their lifestyle. His discussions in *PA* IV. 5 and 7 make this quite clear.)

I think myself that this theory of spontaneous generation is a kind of aberration in Aristotle's thought, which he never properly assimilated and which for most of the time he simply ignores. Time and time again he will insist that man begets man, and he never adds in the same breath that oysters do not beget oysters; this latter point he conveniently forgets (save occasionally, e.g. in *Metaphysics Z*. 9). I began this section with a number of passages in which he very clearly says that the laws of matter must leave a great deal unexplained, and for that reason teleological explanations are essential. Perhaps he *could* have argued, as Nussbaum and Sorabji and Freudenthal suppose that he did argue, that teleological explanations are needed anyway, even if material explanations are available too. (That, after all, seems to be roughly the position that we ourselves are in nowadays.[18]) But for the *most* part that is not what his actual

[17] It is observed by Balme (1962: 96–7) that in the phrase 'spontaneous generation' the word 'spontaneous' is used in a way that conflicts with the definition of spontaneity (τὸ αὐτόματον) in *Phys*. II. 4–6. Lennox (1982) tries to show that there is no real conflict, but I think unsuccessfully. Judson (1991*a*) more plausibly suggests that this is just a different use of the same word.

[18] I mean: we do suppose that the laws of chemistry will by themselves explain just how (given a suitable environment) a fully formed animal will result from the combination of sperm and egg at

practice in biology suggests. For the *most* part it seems that his explanations in terms of what he calls 'necessity' can take one only a small part of the way, and we cannot give *any* account of the rest without invoking something different, namely the purposes of nature. Admittedly, his belief in spontaneous generation conflicts with this position, but it seems to me that, for an overall picture of Aristotle's practice in this area, we do best just to set that belief aside.

In this section I have been considering Aristotle's *practice* in his biological investigations. However, we cannot simply take it for granted that his *theory* of these investigations will match his actual practice. Indeed, there is quite good reason to say that it does not.

4. ARISTOTLE'S TELEOLOGY IN THEORY

We now see rather more clearly why Aristotle insists that matter and form are *causes*, i.e. have a crucial role to play in explanation, and why he also insists that the student of nature must make use of both of these causes. He conceives of his own biological explanations as doing just this. What he is trying to explain here is why animals are as they are, or why they grow and develop as they do, with just these bodily parts and not others. In this enquiry, matter is a cause because one needs to rely upon what I have called 'laws of matter', stating what kind of matter will be produced under this or that condition. Aristotle himself never uses any such phrase as 'a law of matter', largely because he thinks of causes as objects, and supposes that explanations function by pointing to the right objects. (In this he follows Plato.) But the 'necessity' that he so constantly invokes in his explanations surely does refer to something that we would call a law. This 'necessity' he contrasts with 'nature', and very often he writes as though he were personifying Nature, as a kind of benevolent craftsman making the best possible use of the materials available. But presumably he must recognize that this is only a metaphor, and his way of cashing it is to say that what is intended is just *the nature* of whatever kind of animal is in question, which he also identifies with the *form* of the animal. (This he sometimes identifies further as what is stated in the *definition* of the animal, i.e. what it is to be that animal.[19]) This form of the animal is the form that is exemplified in the perfect specimen of the *mature*

conception. But we do not suppose that such chemical goings-on will explain *everything* about animals. In particular, as I observed earlier, a different line of thought is needed to explain why animals are so well adapted to their environments.

[19] Cf. 663^b24 (quoted above): ἡ κατὰ τὸν λόγον φύσις. Gotthelf (1985*a*) has collected and discussed the few passages in *PA* II–IV where Aristotle explicitly assigns some feature to the definition (λόγος) or essence (οὐσία, τί ἦν εἶναι) of an animal. He finds only ten, and many of these concern highly generic groups of animals. Nowhere in *PA* (or anywhere else) does Aristotle actually *give* a definition of any species of animal. For some discussion of this point see e.g. Pellegrin (1985) and Balme (1980).

animal, and it functions as a cause by way of being the goal towards which the animal develops.[20]

As I shall show in a moment, this surely is Aristotle's account of how form functions as cause in the issues he is here concerned with, but in fact the account is quite misleading. Consider again, for example, the question of horns. Why do some animals—say the bull—have horns? On the one hand we have the answer in terms of matter, noted already, but what do we find on the other hand? We do *not* simply find the bare and quite unilluminating answer that having horns is part of what it is to be a bull. Instead Aristotle claims very plausibly that the purpose of horns is that they are weapons for attack or for self-defence. Moreover, he goes on to explain several features of horns as due to this purpose; for example, the most effective place for such a weapon is indeed on the head, where nature has placed it ($663^{a}34$–$^{b}12$); for greater strength, horns always have solid tips, and when—as is usual—they are hollow lower down, they are fitted firmly at the end to a solid piece of bone ($663^{b}15$–20). What has this to do with the 'form' of the animal concerned? Standardly, when Aristotle invokes nature, and its purposes, the explanation that he gives aims to show why it is *good* for the animal to have whatever part is in question (and very often we can add: good for the *survival* of that animal). No doubt Aristotle also thought that it was good for an animal to attain its mature form, and that can not unreasonably be called 'the nature of' that animal. But in fact the personified and benevolent nature that he calls upon so often usually cannot be understood simply as 'the nature of' the animal concerned. It is better understood as what is good for the animal, for that genuinely does play an explanatory role in biology, whereas the animal's nature, or form, simply does not.

Nevertheless, I think it is clear that Aristotle himself thinks of his explanations by means of purpose, or 'that for the sake of which', as ones that invoke form. That is why he tells us, when discussing the causes, that the cause as form and as goal will often coincide. Similarly, equating form with essence, he tells us in the first chapter of *De Partibus Animalium*:

> We should if possible say that because this is what it is to be a man, therefore he has these things; for he cannot be without these parts. Failing that, we should get as near as possible to it: we should either say altogether that it cannot be otherwise, or that it is at least good thus. And these follow. And because he is such a thing, his coming-to-be necessarily happens so and is such. And that is why this part comes to be first, and then this. And this is the way we should speak of everything that is composed naturally. ($640^{a}33$–$^{b}4$, tr. Balme)

The *best* kind of explanation is thus to explain some part by showing that it is implied in the essence. Nearly all the explanations that he actually gives are in

[20] What is in the animal, from the time when its life begins, is a tendency to develop towards this goal. We might wish to regard this tendency as an *efficient* cause, for we are apt to think that final causes cannot actually explain anything if they cannot also be seen as 'acting' via an efficient cause. (Cf. Irwin 1988, ch. 5; Furley 1996; Lennox 2001*b*, ch. 8.) Whether Aristotle would agree with this is not altogether clear, but the point is of no importance here.

fact of the 'second-best' kind: the animal has such-and-such a part because either it could not live at all without it, or not so well, or its presence follows from that of other parts which are independently explained. But these 'second-best' explanations, which genuinely are explanatory, Aristotle here regards as 'as near as possible' to the first kind, which actually are not explanatory at all. Elsewhere he simply writes as though they *are* of the first kind, for example in the later introductory chapter at the beginning of *PA* II. After the passage detailing the three kinds of compositions (646ᵃ13–24, quoted earlier), he continues:

> Now the manner of coming into being (γένεσις) is opposite to that of being (οὐσία), for what comes into being later is prior in nature, and what is last to come into being is first. For the house is not for the sake of the bricks and stones, but they are for the sake of it, and the case is similar with other matter too. That this is the way of things is clear not only from induction, but also from argument. Everything that comes to be does so *out of* something and *towards* something, and from one principle (ἀρχή) to another, that is from the one that begins the movement and already has a certain nature, to some shape or other similar goal (τέλος). For man begets man, and plant begets plant, out of the underlying matter in each case. Thus the matter and the process of coming to be must be first in time, but in definition (λόγος) the essence (οὐσία) and the shape of each thing are first... Hence the matter of the elements must be for the sake of the homoiomerous parts, since they come into being later, and in turn the non-homoiomerous parts are later than them. These latter, whose composition is third, have already reached the goal and the limit to the process. In many cases, the process of coming into being reaches its goal thus [i.e. at the third stage of composition]. (646ᵃ25–ᵇ10)

Although Aristotle's usual word for form (namely εἶδος) is again not actually used in this passage, it is clear that it is form he is speaking of as what comes into being last in time but is first in being (or essence, οὐσία) and in definition. It is said to be a 'principle' (ἀρχή), which evidently here means a cause, and is contrasted with another of Aristotle's four causes, 'the one that begins the movement'. Moreover, it is twice glossed as the *shape* of a thing, and this is a standard way of referring to form in Aristotle. So one is tempted to say that this passage tells us that the detailed explanations of the various parts of animals that are to come will rely on two kinds of cause (two explanatory features), namely matter and form, form also being identified with essence and with goal. (The remaining kind of cause, 'that which begins the movement', is to be discussed later as the main topic of *De Generatione Animalium*.) As I have observed, it would be better to say that these explanations rely on matter and on 'the good', but 'the good' has been assimilated to 'that for the sake of which', which then becomes 'the goal', and this is finally equated with form.

But there is a further, and very puzzling, question raised by this passage: undoubtedly it says that form is a cause, but it does *not* say that matter is. To be sure, matter is mentioned as that *out of which* things come to be (ᵃ31, 34), which is contrasted with that *towards which* they develop, i.e. form. But whereas form is called a 'principle' (ἀρχή) matter is *not*. The passage mentions two 'principles', but they are the form and the *efficient cause*, which (as we recall from *Physics* II. 7, 198^{a}26) has the same form as what it leads to.[21] Moreover, a 'principle' is surely something that comes first, but what we are told about matter is that in the order of nature it comes *last*. It would seem to follow that it cannot be a principle, and hence is not to be counted as one of the causes.

The same implication can be extracted in another way. The matter, we are told, is for the sake of the goal, and not vice versa. Now it is not at all surprising to find Aristotle saying that the homoiomerous parts of the body exist for the sake of the organic parts that are formed from them, but one did not quite expect him to say that the four elements exist for the sake of these homoiomerous parts in turn. Apparently he does quite seriously mean that absolutely *all* matter exists only for the sake of what can be made out of it. Does this not once more suggest that what used to be called the 'material cause' is now no longer being assigned any genuinely independent role? This brings us to a central issue in the interpretation of Aristotle's reflections on how the explanation of living things does actually work.

5. THE ARGUMENT OF *PHYSICS* II.9

This chapter opens with the question whether necessity applies (to natural objects) only 'from a hypothesis' (*ἐξ ὑποθέσεως*) or also in an unqualified way (*ἁπλῶς*). This somewhat obscure dilemma becomes a little clearer as we proceed. People currently think, Aristotle says, that necessity is to be found in cases of coming to be 'as if one held that the wall had come into being of necessity, on the ground that heavy things by nature go downwards and light things to the surface. For that reason the stones are at the bottom as a foundation, the earth[22] above them because it is lighter, and the wood on top as the lightest of all.' This view (though Aristotle does not quite say so) is presumably a view according to which necessity applies 'in an unqualified way'. In any case, he of course rejects this explanation, on the ground that the wall comes into being 'not without these, but also not on account of them (except as matter); rather, it comes into being for the purpose of concealing or protecting' (199^{b}35–200^{a}7). It is the same with other things that have a purpose. They come to be 'not without the things that have a necessary nature, yet not because of these (save as matter), but for the sake of something'. For example, if we ask why a saw is as it is, the answer will run on

[21] The Greek is: *πᾶν γὰρ τὸ γινόμενον ἔκ τινος καὶ εἴς τι ποιεῖται τὴν γένεσιν, καὶ ἀπ' ἀρχῆς ἐπ' ἀρχῆν, ἀπὸ τῆς πρώτης κινούσης καὶ ἐχούσης ἤδη τινὰ φύσιν ἐπί τινα μορφήν* (^{a}31–3). I believe that Lennox (2001*a*) mistranslates. He does not distinguish *ἔκ* from *ἀπό*, and he inserts an 'i.e.' between the first and the second clause, yielding: 'Every generated thing develops from something and into something, i.e. from an origin to an origin, from the primary mover which already has a certain nature to a certain shape.' This translation identifies the matter with the efficient cause, even though the efficient cause is here characterized as 'already having a certain nature'. I see no good reason to accept this identification (*pace* Lewis 1988: 54).

[22] Earth: that is (probably) bricks, made of mud or clay.

these lines: its function is to cut in a certain way, and it cannot do this without teeth of a certain sort, and these can only be made of iron. So 'it is necessary that it should be iron, if it is to be a saw and perform its function (ἔργον). Hence the necessity is from a hypothesis, but not as a goal; for the necessity is in the matter, but that for the sake of which is in the definition' (200^{a}7–15).

This paragraph clears up some obscurities, but it certainly leaves others. First, I think it is clear that when Aristotle speaks of necessity 'from a hypothesis', the relevant hypothesis always concerns an end or goal, and what is necessitated by this is always taken to be matter. The general form of this type of necessity is 'If such-and-such an end, goal, or function is to be realized, there must be matter of such-and-such a kind'. At any rate, this model fits all the examples of necessity 'from a hypothesis' that Aristotle gives, both in this chapter and in the similar chapter *PA* I. 1. Moreover, the latter contains a general characterization of the kind of necessity relevant to the study of nature which, it seems to me, forces this characterization:

> The mode of demonstration and of necessity is different in natural science and in the theoretical sciences . . . [23] For the starting point (ἀρχή) is in the one case what is and in the other what will be. For it is because health, or a man, is such a kind of thing that it is necessary that this should be, or come to be; it is *not* that, because this is or has come to be, it is necessary that that other thing is or will be. (640^{a}1–6)

Our chapter *Physics* II. 9 makes a very similar point in its second paragraph (200^{a}15–30), but it is not so explicit that necessity in nature always starts from what *will* be.[24]

Now Aristotle clearly believes that this necessity 'from a hypothesis' does apply to natural objects (as well as to those, such as a wall or a saw, that are the products of human skill). But the question that he began by raising was: is there *also* an unqualified necessity that applies? This question seems not to be directly addressed in our chapter in the *Physics*, but we do find further information in the similar discussion in *PA* I.1. That passage tells us that in things which exist by nature 'unqualified necessity' applies to what is eternal (e.g. to the behaviour of the heavenly bodies[25]), whereas necessity 'from a hypothesis' applies also to

[23] Since Aristotle standardly counts natural science (ἡ φυσική) as itself one of the theoretical sciences, the contrast intended here must be between natural science and the *other* theoretical sciences. (The similar passage in *Phys.* II. 9, i.e. 200^{a}15–30, contrasts the necessity in natural objects with the necessity in mathematics.)

[24] Charles (1988) is I think alone in supposing that Aristotle's hypothetical necessity includes *both* necessitation by future goals *and* necessitation by antecedent circumstances. He attempts to square this view with the passage just cited by saying that Aristotle is there claiming only that in the natural sciences reasoning must *start* with necessitation by goals. This, he says, is compatible with the view that necessitation by antecedent circumstances may then come in as the reasoning *proceeds* (p. 7). That view (which I discuss later) is a possible one, but one can defend it (as Cooper does) without the implausible suggestion that Aristotle means *both* kinds of necessity to fall under his heading 'necessity from a hypothesis'.

[25] I take the example from *GC* II.11, 337^{b}10–14. (No example is given in *PA*.)

everything that comes into being (639^b21–5).[26] This very strongly suggests, though it does not explicitly say, that 'unqualified necessity' has no application to things that come into being. One would certainly not *expect* such a claim, for a very natural suggestion would be that the study of things that come into being would frequently invoke 'unqualified necessities' of the kind that I have been calling 'laws of matter', laws that Aristotle attempts to explore in *Meteorologica* IV, and so constantly invokes as necessities in *PA* II–IV. An instance is apparently given in his first example here: it would appear to be an unqualified necessity of this sort that heavier objects sink and lighter objects rise, and again that stone is heavier than earth and earth heavier than wood (200^a2–4) (for these are presumably eternal truths). One might at first suppose that in the *Physics* passage he is allowing such necessities some role. Thus in his first example when he says that the wall comes into existence 'not without these' he could be taken as meaning 'not without these *necessities*, e.g. the necessity that heavy objects sink' (200^a5). Again when he generalizes the point, and says of all objects with a purpose that they come into existence 'not without the things that have a necessary nature (τὰ ἀναγκαίαν ἔχοντα τὴν φύσιν)', one could take him as allowing that there is indeed an unqualified necessity in the matter, which is required, but not sufficient in itself. Moreover, this appears to be what he *ought* to mean. For consider again the example of the saw, and suppose that we have worked out that if it is to fulfil its function then it must be made of a material with just such-and-such properties. Very well then, why not make it out of butter? The answer, evidently, is that butter does *not* have the required properties, whereas iron does. But this answer does not draw upon any conditional necessity, conditional upon the hypothesis that there is to be a saw. Rather, it states without qualification what the properties of iron and butter actually are, and this may well be regarded as a *necessary* feature of what iron and butter in fact are. Generally, it would seem that there could not be any 'hypothetical necessities' unless there were *also* some unconditional necessities of the kind just illustrated.

But on reflection it is very difficult to suppose that this is what Aristotle himself means to be saying. Although he opens the chapter by asking whether, as well as hypothetical necessity, there is *also* an unqualified necessity in natural objects, his discussion in fact proceeds as though these were mutually exclusive alternatives. He demonstrates (by using his analogy with skill) that there certainly is hypothetical necessity in nature, and so concludes that *the* necessity in nature is hypothetical. This is already apparent from the concluding sentence of his first paragraph: 'So the necessity is from a hypothesis, and not as a goal; for the necessity is in the matter, whereas that for the sake of which is in the definition' (200^a13–15). Evidently only *one* kind of necessity is being contemplated, and

[26] The 'also' seems to imply that hypothetical necessity applies in *both* cases, which may be somewhat unexpected, but is not impossible. For Aristotle does think that the heavenly bodies move as they do because they are pursuing a goal.

this is said *both* to be 'from a hypothesis' *and* to be 'in the matter'. So I take it that Ross's elucidation (1936, ad loc.) must be correct: Aristotle's meaning is that *what* is necessitated is the matter and not the goal; the necessity is *in* the matter, and the matter has 'a necessary nature' (200^a8–9), *only* in the sense that the matter is necessitated by the goal and not vice versa. The same assumption, that there is only *one* kind of necessity to be found in natural things, is perfectly clear in the next paragraph, where *the* necessity in things that come to be by nature is contrasted with *the* necessity in mathematics. There is not the slightest suggestion that *two* kinds of necessity might play a role in either of these subjects, and again the necessity in nature is clearly identified as hypothetical (200^a15–30). Moreover, we have noted earlier that *PA* 639^b21–640^a6 certainly appears to imply that the *only* kind of necessity that is relevant to the study of things that come to be is hypothetical necessity, and the same implication is explicit in a later passage of *PA* I. 1:

> There are then these two causes, that for the sake of which and that which is of necessity. For many things come about because it is necessary. One might perhaps be puzzled over what kind of necessity is meant by those who say that things happen of necessity, for neither of the two kinds distinguished in the philosophical works could apply. But at least in the case of things that come into being there is the third necessity. For when we say that food is something necessary it is in neither of those ways, but because one cannot exist without food. This is as if from a hypothesis. (642^a1–9)

Unfortunately the reference to 'the philosophical works' (τὰ κατὰ φιλοσοφίαν) is not at all clear. The most obvious passage to point to is the lengthy discussion of necessity in *Metaphysics* Δ. 5, which is briefly recalled at *Metaphysics* Λ. 7, 1072^b11–13. However, this distinguishes *three* kinds of necessity, and one of these *is* hypothetical necessity. (The other two are 'what is forced' and 'what cannot be otherwise, i.e. what is necessary without qualification'.) So commentators have also pointed to *Posterior Analytics* II. 11, 94^b37–95^a3, where indeed just two kinds of necessity are distinguished, namely what a thing does in accordance with its own nature (as a stone falls downwards), and what it is forced to do against its nature (as a stone may be thrown upwards). But here the distinction is only made very briefly in passing, and it would seem on the face of it that *both* the necessities here mentioned *do* apply to things that come into being. So I am inclined to suspect that the reference may be to a lost work.[27] But, however exactly the reference is understood, the overall message of our passage

[27] In two other places Aristotle refers to τὰ κατὰ φιλοσοφίαν. At *Politics* 1282^b19 the reference is almost certainly to *Nicomachean Ethics* V. 3. At *Eudemian Ethics* 1217^b23 it is assumed in Woods's translation (1982) that the reference is to the lost *On Philosophy*, but I think that this is improbable for two reasons: (i) Aristotle standardly refers to this work as τὰ περὶ φιλοσοφίας, not τὰ κατὰ φιλοσοφίαν (ii) what he here says is 'the matter has been discussed in many ways, both in the "exoteric" writings and in τὰ κατὰ φιλοσοφίαν', and this clearly indicates a contrast. What exactly Aristotle means by 'the exoteric writings' is disputed, but I think it is a fair guess that this refers to Aristotle's own published works, in which case they would include the *On Philosophy*, and could not

must be that hypothetical necessity is the *only* necessity that applies to things that come into being.

On the other hand it is clear both that this doctrine is very implausible and that our two texts are not entirely unambiguous on this matter. I began with a passage from *PA* II. 1 which appeared no longer to count matter as a 'cause' (αἰτία) or 'principle' (ἀρχή), but that by itself carries little weight, for matter is called a 'cause' at *Physics* II. 9, 200^{a}30–4, and at *PA* I. 1, 642^{a}14–17, and in the latter passage it is also a 'principle'. The former, however, contains an awkward ambiguity:

Evidently then what is necessary in natural objects is what we call matter, and the changes of matter. And the student of nature should recognize both causes, but more especially the cause for the sake of which. For that for the sake of which is the cause of the matter, while matter is not the cause of the goal. (200^{a}30–4)

We *could* read this passage as saying that, while the *presence* of the matter is necessitated by the goal, still 'the *changes* of matter' are necessitated by *something else*, presumably by what I have been calling 'the laws of matter'. Since the passage goes on to describe the goal as *the* principle (ἀρχή), apparently implying that there is no other, there would be no temptation to interpret it in this way if there were no support from elsewhere. But there is such support, for the *De Partibus Animalium* discussion ends with an example which can only be seen as invoking *two* kinds of necessity:

Exposition should be as follows: breathing exists for the sake of *this*, while *that* comes to be of necessity because of *those*. Necessity signifies sometimes that if there is to be that for the sake of which, these must necessarily be present; and sometimes that this is their state and nature. For the hot necessarily goes out and comes in again when it meets resistance, and the air must flow in; so much is already necessitated . . . (642^{a}32–7, tr. Balme)[28]

Here we clearly have *two* kinds of necessity, *both* said to apply to natural objects and processes. There is the familiar hypothetical necessity, which is here not illustrated. (Presumably it takes some such form as this: since the purpose of breathing is *this*, there must of necessity be such-and-such materials present—some air, and some heat, perhaps.) But there is also *another* necessity, of the kind that we so often find in Aristotle's detailed discussion of particular cases, namely 'that this is the state and nature' (sc. of the materials). (For example, when the heat comes back into the body, the air must flow in with it, for this is the nature

be contrasted with it. If so, then the contrast will be with the unpublished treatises, which include all those that we possess, but could include more as well. (The matter which this text says has been discussed elsewhere is just the problems with Plato's theory of forms. This of course is discussed in very many places, including no doubt the *On Philosophy*; cf. fr. 11, Ross.)

[28] It may be noted that the explanation of breathing given here appears not to be Aristotle's own explanation. At any rate he gives a different account at *De Respiratione* 480^{a}16–^{b}12.

of heat and air.) This second necessity seems not to be 'from a hypothesis', but is apparently 'unqualified'.

How should we conclude? *One* possibility is to suppose that Aristotle never really means to say that the *only* necessity to be found in things that come to be by nature is necessity 'from a hypothesis'. In outline, the explanation will be that he is simply taking it for granted that what I have called 'the laws of matter' do have a role to play in scientific explanations (and do count as necessities). He takes it for granted because it is something that his opponents also accept, and the *main* point that he wishes to make is that these necessities are not by themselves sufficient. We must accept that there is *also* necessitation by future goals, and that this is in fact the more important kind of necessity, for it is where all explanations *start*. This is because scientific explanations must start with definitions, and—at least in the case of living things (which is clearly the main focus of his attention)—the definition of a thing does state the goal at which its development aims. (This is the point that he emphasizes at 200^a15–30.) As the explanation proceeds it may also invoke the 'laws of matter' at some points, and the suggestion is that where our texts appear to rule this out they should be regarded as (admittedly misleading) exaggerations of what he sees as his most important claim. Cooper (1985, repeated in his 1987) is a persuasive advocate of an interpretation along these lines.

The main advantages of this interpretation are obvious: it gives Aristotle a comprehensible position, and one that squares well enough with his actual practice in *PA* II–IV and in *GA*. But really it has very little textual support. As we have seen, almost all of the discussion (both in *Physics* II. 9 and in *PA* I. 1) is most naturally interpreted as claiming that the *only* necessity that has a role to play in this area is necessity 'from a hypothesis', and the only clear counter-example is the final paragraph of *PA* I. 1 (i.e. 642^a31–$^b2$). But at least one commentator (Peck 1961, ad loc.) has said of this paragraph that it 'looks like a displaced note', and 'if it is to remain in the text it would follow after 642^a13 least awkwardly'. When we add (as Peck did not) that the account of breathing that it contains appears not to be one that Aristotle himself endorses, and that its doctrine appears not to harmonize with most of the rest of the chapter, it is clear that its evidence deserves to be treated with some suspicion. So let us turn to the alternative interpretation.

This takes quite seriously the fact that Aristotle certainly does seem to say that necessity 'from a hypothesis' is the only kind of necessity that has a role to play in the explanation of the parts of animals and the generation of animals. One could suppose that this is due to a *confusion* on his part. As we have seen, he describes matter as 'having a necessary nature' (200^a8–9), he says 'the necessity is in the matter' (200^a14), and again 'what is necessary in natural objects is what we call matter, and the changes of matter' (200^a30–2). I have noted that, when the first two of these phrases are taken in context, they must be taken to mean 'the presence of matter (of a certain kind) is necessitated by the goal', and the third may certainly be taken in the same way. But of course the very same phrases *could* be taken to mean 'the behaviour of the matter is necessitated by its *own* nature' (which is clearly what is meant by the very similar phrase at *PA* 640^{b}8–11, describing the position of Aristotle's opponents). It is possible that Aristotle did not clearly distinguish these two meanings. The reasoning that he gives supports only the first, but he may have allowed himself to suppose that this somehow carried the second with it. However, there is also the possibility that, while he sees well enough that the two are distinct, he nevertheless thinks of the first as implying the second, i.e. he thinks that 'the laws of matter' are themselves a consequence of nature's goals. An interpretation along these lines was proposed by Balme in his (1972), and although Balme himself later came to reject it (in his 1987: 285 n.) still I think it is worth exploring. For it seems to me to be a position that Aristotle *could* have believed.

6. EXTREME TELEOLOGY

We may begin by going back to the example of the saw. If there is to be a saw, then there must be a hard, fairly rigid, but not brittle material, and it must be capable of being given an appropriate shape. Let us agree to call such a stuff 'iron'. Then if there is to be a saw there must be iron, but we do not have to cite any separate material necessity that iron should be hard (etc.), since that is now simply a matter of the definition of iron. However, since a saw is not a natural object, there is no necessity that there should be a saw, and so it is not for this reason necessary that there should be iron.[29] By contrast, animals are natural objects, and it is a fundamental fact about the universe that there are animals. So, given this fact, it is necessary that there should be rigid stuff such as bone, liquid stuff such as blood, muscular stuff such as flesh, and so on. Balme continues 'flesh and bone necessitate the action of heat and cold upon foodstuff, which itself necessitates other actions of heat and cold upon earth and water' (i.e. to grow the foodstuffs in question), and so on. The idea is that from the premiss that there are animals capable of perception and locomotion one can deduce first that there must be stuffs with the properties of blood, bone, and flesh, and hence that there must be appropriate processes of nourishment, and hence that there must be plants and the stuffs they are made of, and so finally that there must be the elements earth, air, fire, and water. *Everything* that there is in the world is 'hypothetically necessitated' from the premiss that there are animals of the familiar kinds, living their familiar lives. And it is not only the materials that are necessitated, but also the laws of their interactions. It has to be admitted that, for the sake of exposition, in *PA* II–IV and in *GA* Aristotle does indeed talk of the

[29] It is difficult to see why it is in *any* way necessary that there should be iron, on Balme's account.

'necessary' properties of matter, in contrast to the properties that serve some purpose, but in truth these properties are necessitated only hypothetically, i.e. they are necessary only given the end that there should be animals. But that end in turn—though Aristotle never calls it necessary—is simply a datum. It is a fundamental and not further explicable feature of the universe that it does contain animals. Its other features all follow from this. No doubt, one could not in practice build an explanation of the whole world that began simply from this premiss and no other, but might Aristotle have *thought* that it could be done?

I have remarked earlier that Aristotle does not suppose that all things exist for the sake of *men* in particular. Although he does once allow himself to say this, at the beginning of a work devoted wholly to men (i.e. *Politics* I. 8, 1256^{b}15–22), it is so often contradicted by what he says or implies elsewhere that one can be sure that it is not his considered opinion.[30] I suspect that the position is similar with the passage from *De Partibus Animalium* that Balme relies on. Although it does apparently say that all things exist for the sake of *animals*, it too is very probably an exaggeration. Would it not be more natural for Aristotle to hold that all things exist for the sake of animals *and plants*, these being the two kinds of natural object that he recognizes, other than stuffs? For example, one imagines that he would hold that the purpose of an apple concerns its role in propagating a new apple tree, rather than to be food for animals. And the purpose of the apple tree itself, one would guess,[31] is simply to attain its mature form, and not—say—to provide shade, or food, or wood for other things to use. (Similarly the purpose or goal of a sardine or an earthworm is not, presumably, to be eaten by other animals.) But we can widen the ultimate goals of nature to include plants as well as animals without much disturbing Balme's account of how the *only* necessity to be found in nature is 'hypothetical necessity'.

However, I am inclined to think that we should widen nature's goals yet further. First, let us observe that at one place Aristotle does explicitly say that the elements, and the inorganic compounds that are formed from them, are 'for the sake of something', equating this with the view that they have a 'function' (*ἔργον*) which in turn is a power (*δύναμις*) to act or be acted upon. This occurs in a curious passage from *Meteorologica* IV. 12 (just before the passage quoted earlier), to which interpreters have generally given little attention.[32] As the passage admits, *Meteorologica* IV has concerned itself only with the material nature of these stuffs, for that is relatively much clearer, but they do have a

[30] Sedley (1991) is one of the few who have supposed that this *is* Aristotle's real opinion. There is a response to his arguments in Wardy (1993); see also n. 10 above.

[31] In fact Aristotle several times says that the only *function* (*ἔργον*) of a plant is to reproduce itself (e.g. *GA* 717^{a}22–4, 731^{a}25–6). But he surely cannot have meant this to exhaust its *form*, for otherwise there would be no account of what the form was. (To be self-reproducing is characteristic of all living things, and so cannot by itself distinguish one species from another.)

[32] I think the attitude of Cohen (1996) is typical: in a footnote (p. 143, n. 14) he dismisses the passage with '*de minimis non curat lex*'.

further 'teleological' aspect as well:

We grasp, through the way they come into being, what the homoiomerous parts are constituted from, and their kinds (γένη) and to what kind each belongs. (For they are formed out of the elements, and out of them as matter are formed the whole works of nature.) But while each thing is formed from matter as described, it is, so far as its essence is concerned, given by definition (ἐστι ... κατ' οὐσίαν τῷ λόγῳ). However, this is always more clear in the case of things that come later, and in general those that are as it were instruments and for the sake of something. Thus it is more clear, that a corpse is a man only in name, and similarly the hand of a dead man is a hand only in name, just as a stone flute might be said to be a flute; for these indeed seem to be, as it were, certain instruments. But the point is less clear with flesh and bone, and still less with fire and water. That for the sake of which they are is least clear where there is most matter. If one takes the extremes, matter is nothing other than itself, and essence (οὐσία) is nothing other than definition. Similarly, then, the intermediates correspond to what is nearer to them, since each of them is for the sake of something, and it is not water or fire in any and every state that is water or fire, just as also with flesh and intestine. This applies even more to a face or a hand. But in fact everything is defined by its function (ἔργον) and it is the things that are able to perform their function that genuinely are those things. For example, an eye is an eye if it sees, and an eye that cannot see is an eye only in name ... The same is also true of flesh, though its function is less clear than that of the tongue. Similarly too with fire, though its natural function is perhaps even less clear than that of flesh. Similarly again with the [homoiomerous stuffs?] in plants (τὰ ἐν τοῖς φυτοῖς) and with inorganic things (τὰ ἄψυχα) such as bronze and silver. For everything is what it is by some power (δύναμις) either to act or to be acted upon, as are flesh and sinew. But their definitions are not accurate [sc. accurately known?], so it is not easy to discern when they apply and when they do not. (389^b25–390^a21)

Apparently Aristotle envisages that just as a dead hand should not really be called a hand, since it can no longer fulfil the function of a hand, so also there are states of fire and water, of bronze and silver, in which they do not deserve to be called fire or water, bronze or silver. But, he says, it is not clear quite what these states are, for it is not clear what the function of these stuffs in fact is.

Now from the fact that Aristotle here says that various functions are not clear, we should not infer that he is himself doubtful about them. The passage is designed to lead into his discussion of the parts of animals, and in that discussion he will in fact explain the functions of all the parts of the body that he here mentions. These functions, then, are said not to be clear because, in the order that Aristotle designed for his works on nature, we have not yet reached the discussion of them. But it is not so easy to understand why Aristotle apparently treats the proper functions of the elements too as an open question, for surely that has been settled already in the *De Caelo*?

The notion of a thing's function is not very much used in the *De Caelo*, but it does twice occur in just the way one might expect, identifying the proper function of an element with its natural motion. Thus in book II, speaking of the 'divine' fifth element (αἰθήρ), Aristotle says that 'everything which has a function exists for the sake of that function' (286^{a}8–9), and proceeds to specify the function of this element as its eternal circular motion. Then later, when he turns to a full account of the four sublunary elements, he prefaces this by remarking that the most important differentiae of bodies are in their attributes (πάθη), their functions (ἔργα), and their powers (δυνάμεις), and adds that every natural body has these (307^{b}20–3). So the ensuing account of the natural motions and natural places of the four elements, and of their heaviness and lightness both absolutely and relative to one another, is surely meant to include an account of their proper functioning. If we may assume an identification between 'function' and 'form', this is clear, since we are told that, for an element, 'being moved towards its own place is being moved towards its own form' (310^{a}34–^{b}1). (Compare also the close connection between the motion of the elements and their essence (οὐσία), indicated at 310^{b}31–311^{a}2.) Why should we not say, then, that the purpose, function, or goal of an element has already been made perfectly clear: it is to move to, and rest in, its natural place? And this also, of course, is its definition.

On the one hand one might reply that this can hardly be what *Meteorologica* IV. 12 would accept as the purpose or function of an element, both because such an account would seem (to Aristotle) to be perfectly clear, and because it would not seem to leave us room to say that there are states of fire and water in which they are not capable of fulfilling their function, and are fire and water 'only in name'. If we accept this verdict, then apparently we must say that in this passage Aristotle is envisaging some further purpose that the elements serve, and this presumably must be—as implied by *PA* II. 1, 646^{b}5–8, and as Balme suggested—the contribution that they make to the lives of animals (and possibly plants). On the other hand we could I think meet the difficulty in this way. Suppose that I pick up a handful of what, indisputably, we call 'earth'. This, then, is certainly earth in name. Suppose also that I then release this stuff, and observe that it shows the appropriate tendency to seek its natural place at the centre of things. Does it follow that it genuinely is earth? The answer is evidently 'no'. The most that follows is that it contains a good deal of what is genuinely earth, enough in fact to counteract the natural tendencies of any water, air, or fire that it also contains. That is, the definition of the element earth, though in a way it is perfectly clear, will not in practice allow us to identify any pure specimens of the stuff. The same obviously applies to fire, air, and water. Perhaps, then, *this* is the problem that leads Aristotle to imply that, even in the case of the elements, their definitions are not 'clear' or not 'accurate', so that 'it is not easy to discern when they apply and when they do not'.

On this interpretation, the goal or purpose of an element is, as it were, its *own* goal, namely to seek its natural place and to remain there. Just as the goal of each animal is fully to realize its *own* form, independently of the purposes of other animals, and similarly (I suggest) with plants, so too we should recognize that even inanimate matter has goals of its own. There are *many* goals in nature. Seeing that the organic compounds formed from the elements can very clearly be assigned goals too (derived from their contribution to the animals—and plants?—formed from them), Aristotle is led to extend the hypothesis to inorganic compounds, such as bronze and silver, as well. So far as I am aware, he never makes any suggestion as to what these goals might be. He may conceive them, by analogy with organic compounds, as derived from the contribution that these stuffs make in some larger scheme of things. (But, if so, I suspect that the larger scheme of things is the ecology of the sublunary sphere as a whole, rather than the needs of animals and plants in particular.) On the other hand I think it more likely that he conceived them, by analogy with the goals of the elements, as 'their own' goals. Thus we humans cook things in order to change them into a new form, which serves our purposes. So nature too will 'cook' things for its purposes, where this 'nature' is then taken to be *the nature of* whatever stuff results from the cooking. (Compare *Meteorologica*, IV, $379^{b}18$–30, $381^{a}1$–4.) Thus the formation of stone, gold, diamonds, and so on will occur 'for the sake of' the stone, gold, and diamonds that result. Aristotle's profound conviction that 'nature does nothing in vain' must evidently dispose him to think that there must be purposes absolutely everywhere in nature.

I add this conjecture for why Aristotle might have thought that even inorganic stuffs have a 'purpose' of their own. It has been observed that from an Aristotelian perspective there is some difficulty in explaining the cohesion of any compound body,[33] and when Aristotle notes the point himself he turns to form to provide the answer. Thus at *De Anima* II. 4 he says that Empedocles tried to explain the growth of plants by saying that the earthy matter in their roots naturally tends downwards, while the fiery matter in their upper parts has the opposite tendency. To this he fairly objects that if this is the full explanation then one would expect the roots and the upper parts simply to separate from one another, each travelling in the direction natural to it. The fact that they do not do so, he suggests, is due to the 'soul' of the plant, which keeps it together as a single thing ($415^{b}28$–$416^{a}18$). Now an inorganic compound, such as a metal, evidently does not have a 'soul' (since it is not a living thing), but it does have a *form*, which might fulfil the same task of holding together the earth and the water that compose it. But the form (i.e. soul) of a living thing Aristotle identifies with its goal, and he seems to think that the tendency towards a goal has an important role in explaining how the soul (i.e. form) can fulfil the various tasks assigned to it. So, by analogy, if the form of an inorganic compound has a similar task to perform, it too must be because the stuff is tending towards some goal.

Well, that is a mere speculation, and I know of no textual support for it. But it certainly is true that in one place, i.e. at the end of *Meteorologica* IV, Aristotle claims that *everything* which exists by nature has its own 'function', which is to

[33] e.g. Gill (1989, ch. 7) and Freudenthal (1995, ch. 1, sect. 1.1). (Freudenthal supposes that Aristotle wishes to invoke *pneuma* as the solution, but this surely would not apply to inorganic stuffs.)

say that there is something which it is *for*. In this mood he could easily be led to think that all of its behaviour is explained by what it is for, so that what may be regarded as 'laws of matter' are always to be derived from these natural 'goals'. That is, he might really have meant, what he apparently says in *Physics* II. 9 (and in *PA* I.1), that the *only* necessity in nature is 'from a hypothesis'. This is because absolutely *everything* that happens regularly in nature serves some purpose, and happens because it serves that purpose. To support such a claim one must, of course, posit many more purposes in nature than we find reasonable, but apparently that is what Aristotle does, at least on some occasions.

As I said earlier about his theory of spontaneous generation, I do not imagine that this line of thought ever played a dominant role in his thinking. The theory of spontaneous generation was one which he felt had to be accepted, in the light of what he took to be the observed facts about some peculiar creatures such as shellfish, but for the most part he shut his eyes to what it implied about the efficiency of mere laws of matter, because this was so very much out of line with all his other thoughts about the importance for living things of their inherited form. In the opposite direction, I take it that he was sometimes led to exaggerate what he saw as the fundamental role of form, i.e. as function or goal, and the corresponding claim that absolutely everything that happens in nature happens for some purpose. But for the most part he steers a more sober course, which gives an independent role both to form and to matter. This applies both to the vast majority of his biological writings, where he is engaged in the search for effective explanations, and to most of his more general reflections on metaphysical issues, where again it is allowed that each has a distinct role to play. So in my final section it is this 'more sober' approach that I aim to evaluate.

7. ARISTOTLE'S THEORY OF EXPLANATION

Aristotle's approach to the explanation of what happens in nature is heavily blinkered in two ways.

First, he focuses on what he calls natural *objects*, and has a rather restricted view even of these. He recognizes a variety of stuffs, i.e. the ultimately underlying ('prime') matter, its four basic forms (earth, water, air, fire), and the various compounds that can be formed from these, both non-living (bronze and silver, wine and vinegar) and living (wood and leaf, flesh and bone). No doubt we can say that in principle all stuffs are here included. But what of the particular things that can be made of them? His standard account mentions only living things and their parts. For example, he does not mention such ordinary objects as pebbles or snowflakes, hills or rivers, waves or whirlwinds, and so on and on without end. Yet these surely are things that exist 'by nature', and also have a 'nature' of their own, not entirely to be explained by the stuff they are made of. They seem to be proper objects of a physicist's attention. But more important than this is that he concentrates on explaining things about his favoured objects (how they come into being, why they grow as they do, why they act as they do). He does not think more generally of the explanation of natural *phenomena* (for example, light or heat or magnetism), but concentrates always upon his favoured *objects*.[34] This point is connected to the second:

He puts himself into a straitjacket right from the start, when he assumes that the explanations he is seeking must invoke either the matter or the form of these objects, and simply never considers any third alternative. This is to some extent due to his somewhat naive understanding of the notion of a 'cause', for he is apt to think of the question 'What is the cause?' on the model of 'Who is the cause?', and so as asking for some responsible agent. (In this he follows Plato.) Consequently he does not see scientific explanations as we do, as invoking what we call *laws* of nature, but as pointing to *things* which explain.[35] This is how he is thinking when he claims that the 'nature' of a natural object is both the cause of its natural behaviour and something within it. I take a very simple example to show how inadequate this is: Why is it that, when one lets go of a clod of earth, it falls downwards? Aristotle says that the clod must have within itself a certain place-seeking tendency. We say rather that this is due partly to something that is perhaps 'within it', i.e. it has a certain mass, but also due partly to something that is certainly not 'within it', i.e. the nearby presence of the earth itself with its very much larger mass, and that the explanation relies also on something which cannot be thought to be 'within' any object at all, namely the Newtonian law of gravitation. Of course, one can hardly reproach Aristotle for failing to be Newton, but one can fairly say that if only he had thought a bit more about what really does go into the answer to a question 'Why?' he should have hesitated before supposing that the answer must always point to something which can be located in some object. Since he does not hesitate over this, one can see why he could easily think that the 'cause' of the natural behaviour of a natural object is within that object itself. But, even so, I do not see why this led him to suppose that it must be either the matter or the form of that object, for (on a natural view of what counts as 'matter' and what counts as 'form') there is no antecedent reason to suppose that these must exhaust all the properties of an object that may be causally relevant. Nevertheless, Aristotle does suppose it, and thus puts on a straitjacket of his own making.

Let us come back to the question of 'laws' of nature. While Aristotle does not think of these as forming even part of an explanation (i.e. an answer to a question 'Why?'), what he says does of course presuppose their existence. To continue with my simple example, his four elements are characterized in two ways, both in

[34] I am speaking of his programmatic descriptions of what 'nature' is, not of his own scientific practice, which provides numerous counter-examples; e.g. almost all of *Phys.* IV–VIII (on space, time, motion, etc.), *Cael.* I–II (on astronomy), *Meteor.* I–III (on various natural phenomena).

[35] Again, I am speaking of his theory, not of his practice.

terms of their natural motions and in terms of the basic 'opposites', hot and cold, wet and dry. He must therefore be supposing that there are general laws connecting these two, e.g. that whatever is purely hot and purely dry (i.e. fire) will naturally move upwards, and that whatever is purely cold and purely dry (i.e. earth) will naturally move downwards.[36] He is also committed to various further laws in what he says about the effects of heating and cooling, and how they may generate a new kind of stuff from what is already present. I shall continue to refer to these as 'the laws of matter'. So what one might expect to find, when paraphrasing Aristotle's own thought into our way of looking at things, is that he invokes two kinds of laws, both 'laws of matter' and 'laws of form'. I think that this is essentially right, but it does need some qualification.

Setting aside the extreme teleology discussed in my previous section, let us consider something more like Aristotle's actual practice in biology, which allows that the laws of matter do have an autonomous role to play, but insists that they are not by themselves sufficient to explain the phenomena. For, at any rate with the parts of animals (and plants), we also need to make use of the idea that the thing has a function, which is to say that it contributes in some suitable way to a goal, end, or purpose. That is what explains why it is shaped and structured as it is (e.g. in the case of teeth or horns), or why it does what it does (e.g. in the case of eyes or stomachs). And how, exactly, does this explanation work? Well, the main thing is that it shows that it is *good* for the animal (or plant) in question to have such a part, doing what it does.

The notion of goodness is crucial. Aristotle leans heavily on the analogy that he sees between instruments designed and made by men and those 'instruments' that are as it were 'designed and made by nature'. Now an instrument has a function because it is designed and used to perform that function, and it is used in this way because men desire what results from so using it; and what they desire they think good. In the case of 'nature's instruments' there is not really any such person as Nature (or God) who has desires and thinks things good, but there is something that counts as good, namely what is good for the animal or plant concerned. So this is the point of the analogy. To assimilate Aristotle's thought to ours, we might say that what is being counted as 'good' is the survival of the species, but Aristotle—at least usually[37]—is thinking of the good of the individual member of the species: it is good for it to live successfully the kind of life that is typical of that species. That is why we can explain why a given animal has a certain part by showing how that part contributes to the good life of the animal as a whole, for it shows that it is good for the animal to have that part. (We cannot give functional explanations of parts which do no good, e.g. the appendix in man. Practically all that the appendix does, in man, is to give one appendicitis.

[36] I add incidentally that there seems to be no saying which of these two characterizations gives the 'form', and which the 'matter', of the elements. The problem generalizes to all stuffs. So in their case there is no saying whether (on Aristotle's own principles) it is their 'matter' or their 'form' that is their 'nature'. The point is noted by Ackrill (1972–3: 132–3).

[37] See n. 31 above.

Since this is not good, we say that in man the appendix has no function, and its presence must be explained in some other way.)

So there is a general law that is being invoked here, which one might call a 'law of goodness', roughly thus: whatever parts a living thing needs, in order to live a life that is good for it, will for that very reason tend to be present in it (and therefore will grow as it grows).[38] The law is limited in its application, of course, by the fact that the 'laws of matter' will only permit some kinds of parts to develop, and not others. (For example, however useful it might be for a man to have a metal skull, the laws of matter do not permit metal to be 'grown' from what men can eat.) But I think it is fair to say that it is some such law of goodness, working together with the laws of matter, that lies behind Aristotle's practice in explaining the 'nature' of his favourite 'natural objects', i.e. living things. Moreover, it is for him a law that is basic and irreducible. This seems a strange position to us, yet it is difficult to see what alternative was available in Aristotle's day, especially given his rejection of a creating God.

But it was a mistake to suppose that this law cites as 'cause' something that can be called the 'form' of an object, and is 'in' that object. We surely cannot construe what is good for an animal as something that is 'in' that animal, in the same kind of way as the shape of a bronze statue may be thought of as 'in' that statue. Nevertheless, I think that Aristotle does try to hang on to his initial assumptions, in this way. The law cites as a goal a life that is good for the animal in question. Dropping the notion of goodness, let us just say that it cites the animal's life as a goal. But now we may identify life and soul, and as I read *De Anima* II. 1 that is just what Aristotle's 'official' account of the soul is trying to do. Yet at the same time he clearly operates with a more ordinary conception of the soul, by which it is the *cause* of life, rather than identical with it.[39] So it comes about that the 'law of goodness' is transformed into the thought that the 'cause' which explains why a living thing grows and develops as it does is its soul, and (on the ordinary conception) this can certainly be construed as something 'within' it. We need only add that the soul of a living thing may be counted as its 'form', and everything is now reconciled.

Needless to say, I do not think that *we* should regard this reconciliation as successful.

[38] This formulation is perhaps stronger than it should be. Sorabji (1980: 158) offers this weaker version of a 'teleological rule' to which Aristotle is apparently committed: 'the organs [of a living creature] will usually do some good'.

[39] I have offered a *brief* account of this ambiguous conception in my (1994: 141–4).

5

Aristotle's Theory of Form

I have argued in Essay 3 that Aristotle's theory of matter was a good theory. Despite what some critics have said, it is internally coherent, and a rational way of explaining the phenomena. Moreover, it lasted for centuries. Naturally there were modifications as time went on, but the basic idea survived in science until (say) the beginning of the twentieth century, and it survives in common sense even today. By contrast, his theory of form was a complete failure. It never became part of common sense, and though some aspects of it lingered on in science (i.e. in biology) until (say) Darwin, they never deserved to.

The basic reason is that Aristotle soon came to think of form as contrasted with matter. Admittedly, it seems that this was not always so, for in the so-called Logical Works (which I take to be early) this contrast does not appear, and while the notion of form used there is somewhat obscure still it does appear to be consistent. But once the contrast became firmly established in his mind, it was the notion of form that suffered. For it had to absorb *everything* that could not be put down to matter, and this made the concept quite unworkable. In this essay I will bring together and compare the many different roles that Aristotle assigns to forms, and so make clear that *nothing* could fulfil them all.

Since the aim is to set out and survey the broad picture, I shall sometimes have to skate rather quickly over problems of interpretation affecting this or that particular passage. I shall also have to exclude the *Metaphysics* from my review, for the problems and difficulties that arise there would simply be overwhelming.[1] So in effect this is a survey of the notion of form as it appears in Aristotle's Physical Works (including the *De Anima*), though I do begin with a brief remark on the Logical Works, which is where the notion is first found. Taking what is obviously an over-simple view of the chronology, one might perhaps say that my aim is to set out the various notions of form that Aristotle brings with him when he starts to work on the *Metaphysics*.

[1] I have tried to tackle the problems in books *Z* and *H* of the *Metaphysics* in my (1994). For some surveys of what happens in those books see pp. 172–5, 185–90, 204–7, 236–7, 287–90.

1. THE LOGICAL WORKS: FORM AS SPECIES?

Those who read Aristotle in English may be surprised to hear that the notion of form occurs at all in the Logical Works. This is because the word εἶδος, which is the word that is standardly translated as 'form' elsewhere, is in the Logical Works usually rendered not as 'form' but as 'species'. But it is important to remember that it is the same Greek word in each case. In the Logical Works 'species' seems a fair translation, for at any rate the examples that Aristotle gives could fairly be called 'species' in our vocabulary. His basic scheme is that there are on the one hand particular things—e.g. particular men, or horses, or trees—and on the other their 'forms' ('species'), and their more general 'types' ('genera', γένη). Thus *animal* is a general type (genus), under which fall many different forms (species), such as *man*, *horse*, *sheep*, and so on. These are all 'universals', but they have as particular instances such things as Socrates, Callias, and Bucephalus. The genus is thought of as 'divisible' into its various species, and these may sometimes be further divided into subspecies, but in this division there is a lowest level of species that are not further divisible in the same way. It is this lowest level that Aristotle is usually thinking of when he uses the notion of a species/form. For example, the species/form *man* does not divide into male and female, or into white men and black men, or in any other way. It is *man* (i.e. human being) that is the form/species, and these further subdivisions are not called forms/species. Why not? Aristotle never really explains.

On his account there are various (simple) things that we can say about a particular individual such as Socrates; we can say that he is a man, is pale, is short, is a husband, is in the Lyceum, and so on. But only the first of these says *what he is* (τί ἐστι), whereas the others assign to him a quality, or a quantity, or a relation, or a location, or whatever. In short, there are different kinds of predicates that apply to a thing, but one and only one of these will give its form (species), i.e. will say *what it is*. (It is admitted that to give the genus, e.g. 'It is an animal', is to provide a *partial* account of what it is. It is usually—but not everywhere—accepted that to give a relevant differentia within that genus, e.g. 'It is two-footed', is also to provide a partial account.) What is so special, then, about the one predicate which does (fully) say what the subject is? What distinguishes it from all the many other things that we may say about it? We do not find, in what Aristotle says himself, any very clear answer.[2] What we do find is a rather unhelpful variety of special idioms which Aristotle uses to mark the special case where the predication tells us what the subject is. Thus in the *Categories* he says that in this case what is predicated is 'said of' the subject rather than 'present in' it, and elsewhere that in this case what is predicated gives the 'being' (or essence) of the subject (τὸ εἶναι, τὸ τί ἦν εἶναι, ἡ οὐσία), or again that it gives 'just what the subject is' (*X* ἐστιν ὅπερ *Y*), or what it is 'in its own right' (καθ' αὑτό). To mark this use of the

[2] I have discussed this question at some length in my (2004), so I do not pursue it further here.

notion of 'form', let us fix on the idea that the form of an object is here taken to be its *essence*. This is merely a terminological stipulation, for I am not supposing that the word 'essence' already has a well-determined meaning. (When the subject is universal, to give its essence is to give its definition; but our concern is the case where the subject is a particular, and particulars have no definition.) Our only real guide so far is the almost unanimous view of translators, that the word 'form' in the Logical Works is to be rendered as 'species', for this both fits Aristotle's examples and has the right contrast with 'genus'.

If one asks a modern biologist what it is that defines a 'species', one might meet this kind of reply: members of the same species interbreed with one another (in the wild), to produce fertile offspring. There is no hint of this in anything that Aristotle himself says. I would not wish to conclude that there was no thought of it in his mind either, for he was himself a biologist, and may perhaps have been influenced by such considerations. (It would explain, for example, why he is so firmly convinced that all men—Greeks and barbarians, white men and black men, male and female—are of the same species.) But it does not explain why he should think that a thing's species (or form) specifies *what it is* in a way that no other predicate does. The best answer to this question seems to me to be an entirely modern suggestion, due essentially to Geach, but since elaborated by Wiggins and others:[3] the form (or species) of a thing somehow embodies the *criterion of identity* appropriate to it. Hence to know the 'form' of a thing is just to know what counts as *the same* thing, as it changes and develops over time. (To use a favourite illustration, we must distinguish between 'the same river' and 'the same water', for Heraclitus stepped often into the first but never into the second.) Since all of Aristotle's primary substances are things that endure through time, the notion of a 'criterion of identity' is highly relevant to them, and the suggestion that his notion of a thing's 'form' is best seen as fulfilling just this role seems to me very attractive. Nevertheless, I shall not pursue this suggestion here, both because there is absolutely no hint in his text that 'form' is supposed to fulfil this role,[4] and because the notion of a 'criterion of identity' is by no means so straightforward as it may at first seem to be. But I recommend the idea: it goes some way towards explaining quite a lot of what Aristotle does say, in the Logical Works, of the importance of his special notion of *what a thing is*.[5]

2. THE *PHYSICS*: FORM VERSUS MATTER

In *Physics* I. 7 Aristotle gives us his own account of how change is to be analysed, and this introduces us to his contrast between matter and form. But in this

[3] Geach (1962, esp. sects. 31–4, 91–2); Wiggins, principally (1967, 1980).

[4] There is even a contrary indication. As I observed in Essay 3, Aristotle insists that (prime) matter has no specific form (no essence) at the same time as insisting that its role is to endure through time as 'the same matter'.

[5] Again, see my (2004) for some further discussion.

account the word 'form' (εἶδος) is used in a quite unexpected way, to characterize any property whatever that a thing may come to acquire by changing. This is by no means limited to the properties which define a species—e.g. his own main example in I. 7 is of a man becoming 'musical'[6]—and it is a much wider use of the word than is usual elsewhere.[7] While we may note it as an instance of how Aristotle can on occasion stretch the meaning of this word, still I think we do best to begin just by setting it aside as a temporary aberration. At any rate, in *Physics* II the word is restricted to a much narrower use, more closely related to that of the Logical Works, though—as we shall see—not identical. One may perhaps think of the connection in this way: *one* kind of change is where a new substance comes into being, as when a seed becomes a tree, bricks are built into a house, or wine turns to vinegar (190^b1–10). In changes of this kind the property acquired is what is usually called a 'form' (a 'substantial form', as commentators often say). But in other changes the same substance remains, and acquires a new property in some other category—e.g. a new quality or quantity or location[8]—and these properties are not usually called 'forms'. Let us press on to what *Physics* II has to say.

The word 'physics' means 'the study of what is natural, of what exists by nature', and *Physics* II. 1 opens with Aristotle's account of which objects do count as 'natural' objects. Wishing to contrast them mainly with objects that are manufactured, he explains them as those that have within themselves a cause of change or stability (192^b8–33), and at once adds that they must all be substances (192^b33–4). He then proceeds to raise the question whether the 'nature' of such a natural substance should be understood to be its matter or its form, and gives some arguments on each side (193^a9–18). His own response in the next chapter is that both contribute to a thing's 'nature', and consequently the physicist should pay attention to both (194^a12–$^b15$). In the course of this he mentions that the earlier physicists erred by concentrating entirely on matter, though it is admitted that Empedocles and Democritus did also touch upon form to some extent (194^a18–21).[9] How should we understand the contrast between matter and form in this context? There is no explicit elucidation in either case.

[6] 'Musical' is a transliteration of Aristotle's word μουσικός, and I suspect not a very good translation. His meaning is probably broader, i.e. 'one who is educated in the works of the Muses'; these include poetry and drama as well as music, and nowadays we would probably include all forms of literature.

[7] But the wide use recurs in several places in the *Metaphysics*, e.g. *Z*. 7, 1032^a26–$^b6$; *Z*. 11, 1037^a30–$^b4$; *H*. 1, 1042^a32–$^b8$; *H*. 6, 1045^b7–23. It also recurs in the theory of the *De Anima* that what is perceived or thought is always a form (e.g. 424^a17–25, 429^a13–29).

[8] According to *Physics* V. 2 this 'e.g.' should be 'i.e.', for there Aristotle argues that change does not take place in predicates of any other category.

[9] I presume that Empedocles touched upon form when he specified the ratio or proportion in which his various elements combined to make particular kinds of stuff (e.g. blood and bone; frs. 96, 98 DK). Democritus touched upon form when he spoke of the different shapes and arrangements of atoms (cf. 184^b20–2).

The arguments that he gives for supposing that a thing's matter contributes to its nature mention both the 'proximate' matter of the thing (as wood is the matter of a bed, or bronze of a statue, 193^{a}9–17) and the more 'ultimate' matter which may go to constitute this in turn (e.g. earth and water, perhaps, 193^{a}17–28). He apparently offers no opinion on which of these is the more plausible candidate, but one guesses that his answer would be 'both'. The arguments that he gives in favour of form seem to introduce a variety of considerations. Before coming to these, let us note that the word 'form' is immediately glossed as 'the shape (μορφή) and the form (εἶδος) as given by the definition (λόγος)' (193^{a}30–1; similarly 'the form as given by the definition', 193^{b}1–2; and 'the shape and form, which is not something separable except in definition', 193^{b}4–5). It is plausible to suppose that form and shape are initially equated because Aristotle still has in mind the bed and the statue (^{a}34–5), but since these are not natural objects he soon switches to flesh and bone (^{a}36–^{b}3). He is apparently thinking of these as *stuffs*, and so not actually defined by their shape, but rather (one presumes[10]) by the ratio or proportion between the different elements—earth, water, air, fire—of which they are composed. This warns us not to take too literally the apparent equation between form and shape (though it is at once repeated, 193^{b}4).

As for *why* it is form (so explained) that has a better title than matter to be a thing's 'nature', one can only say that it is difficult to extract any good sense from Aristotle's first argument (193^{a}31–^{b}6), but perhaps a sympathetic reconstruction of his intentions would be this. An account of the 'nature' of a thing should also be an account of what it is,[11] but to specify only its matter is to give only a very partial account; the form is needed too. (Perhaps we should add: to specify only the form but not the matter—if that is possible—is *also* to give only a partial account, so both are needed.) The next argument rather obscurely connects form with what something is 'actually' (ἐντελεχείᾳ) as opposed to 'potentially' (δυνάμει, 193^{b}6–8). This is a contrast which figures largely in the *Metaphysics*, and for present purposes I shall have to leave it on one side.[12] The third argument is apparently merely *ad hominem*: *if* (as Antiphon had argued) the 'nature' of a bed must be its matter, because if you planted a bed, and it began to grow, then it would give rise to wood, but not to a bed, *then* equally the 'nature' of a man must be his form, since a man gives rise to (i.e. begets) a man (193^{b}8–12). It

[10] See the previous note on Empedocles. But Aristotle also thinks of organic stuffs such as flesh and bone as having *functions*, which is a further aspect of his notion of form that I shall come to shortly. (And at *Meteor.* IV. 12 he apparently extends this idea to inorganic stuffs too. See Essay 4, pp. 71–4.)

[11] A thing's 'nature' is casually identified with its 'substance' (i.e. essence, οὐσία) at 193^{a}9–10.

[12] Aristotle's use of this contrast can sometimes seem very muddled, for example when he appears to claim that since a given piece of matter is everything potentially it cannot ever be anything actually. Also, the notion of potentiality cannot really *explain* anything at all, though Aristotle frequently writes as though it does. For some explanation of these claims see my (1994: 225–6, 251, 283–4, 289).

is not clear whether Aristotle commits himself to the 'if'-clause, but—as we shall see—he does take it to be an important point about form that man begets man. Finally, Aristotle argues that a growing thing is not what it grows from, but what it grows into, and that this is its form (193^b12–18). Again, it is not at once clear why this is supposed to show that the form of the thing is (part of) its 'nature',[13] but there is much more to come on this general theme, so I at once press on to what comes next.

This is the celebrated doctrine of 'the four causes' in II. 3 and II. 7. Although Aristotle quite often insists that a 'cause' is always the answer to a question 'Why?' (194^b16–20; 198^a14–16, $^a23$, $^a31$, $^b5$), still neither his 'material cause' nor his 'formal cause' are here supplied with any 'why'-questions to answer. On the contrary, the material cause appears to answer the question 'What is it made of?' or 'What was it made from?' (194^b23–6, 195^a16–19),[14] and the formal cause appears to answer the question 'What is it?', understood in Aristotle's loaded way as asking for an essence or definition (194^b26–9, 195^a20, 198^a16–18, 32). The form is also somewhat oddly glossed as 'the paradigm' (194^b26)—a Platonic word for Platonic forms—and the one example that we are given is of a ratio or proportion (the formal cause of the octave is 2:1, 194^b28). One wishes that Aristotle had been more generous with his examples. Finally I add, because it will become relevant in a moment, that Aristotle's examples of the final cause do answer suitable 'why'-questions, for they are examples of human actions being explained by the agent's purpose (194^b32–4, 198^a20). It becomes abundantly clear from chapters 8 and 9 that Aristotle also thinks that a natural process may be explained by its purpose, though in this case the 'purpose' is not that of any conscious agent. He also appears to attribute a 'purpose'—i.e. a 'goal' (*τέλος*) or 'that for the sake of which' (*οὗ ἕνεκα*)—to a natural *object*, such as an animal. At any rate, this is surely the implication of a passage which is crucial for understanding Aristotle's notion of form: 'Three [of the four causes] often coincide. For what a thing is is the same as what it is for the sake of, and that from which the change first began is the same as these in form; for man generates man' (198^a24–7). Thus man has a form (what man is), and also a purpose (what man is for), and these are the same; moreover, the efficient cause of a man (what starts off his existence, i.e. the father) is once more something of the same form. While it is left somewhat vague quite how 'often' this triple coincidence occurs, it presumably is intended to apply at least to all living things, which are Aristotle's primary examples of substances.[15] Let us now briefly review these various aspects of what Aristotle calls 'form'.

[13] There is a pun on the word *φύσις*, as meaning both 'nature' and 'growth', but presumably the argument is not intended to rely on this.

[14] I shall set aside, as irrelevant to Aristotle's main thought, the odd suggestion that in an argument the premisses are the 'material cause' of the conclusion (195^a18). This appears to be a reminiscence of the different classification of 'causes' at *Posterior Analytics* 94^a20–4.

[15] It seems probable that Aristotle would wish to identify the form and the purpose of many manufactured objects, e.g. a house (199^a12), a ship (199^b29), a wall (200^a1), a saw (200^a10). At

First, the notion of form is still being explained, as it was in the Logical Works, by a special connection with the question 'What is it?' But secondly, form is also now contrasted with matter in a way which makes this connection quite inappropriate. For in the Logical Works Aristotle primarily thinks of his question 'What is it?' as asked of what he there calls primary substances[16]—e.g. a particular man, or horse, or oak tree—and as generating answers such as 'a man', 'a horse', 'an oak tree'. Yet these answers specify what he now regards as a *compound*, consisting of a certain kind of form *in* a certain kind of matter, and they do not advert only to what he now calls the form of the thing. The point is easily illustrated if we may take a simple example where form really can be regarded just as shape, as with a wooden bed or a bronze statue. The correct answer to a 'What is it?' question in these cases would surely be 'a bed' or 'a statue'; it would *not* be a description simply of the *shape* of the object. Yet a new shape is indeed what the pre-existing matter acquires when such an object is created, and so seems to be a good candidate for what counts as its form. (Of course one might also say that the matter in this new shape comes to acquire a new purpose—a point I shall take up shortly—and it acquires much else too, which no one would think of as contributing to its 'form', e.g. an increased financial value.) We might try to harmonize these points by saying that the form of an object will provide part, but only part, of the answer to 'What is it?', since its matter will have to be mentioned too. But I should also note that many people have urged that it is what Aristotle calls a thing's form, rather than its matter, that has a more important role to play in the criterion of identity appropriate to it. (For living things change their matter over time, and the same seems to apply to artefacts, at least to some extent.) This is one of the attractions of the suggestion that Aristotle's 'What is it?' question is best understood as asking after criteria of identity.

Another aspect of form that is first[17] introduced in *Physics* II is that a thing's form is called a 'cause'. It is plausible to suppose that this theme enters right at the beginning, when 'natural' objects are characterized as those that have an inner cause of their (natural) behaviour, and then their 'nature' is said to consist of their matter and their form. For it is plausible to suppose that Aristotle means the 'nature' of a natural object to *be* just that inner cause of its natural behaviour.

Metaphysics Z. 7, 1032^{a}26–^{b}15 he also claims that the efficient cause of such objects is again the same in form, for the efficient cause is the thought in the mind of the maker. I observe incidentally that in *Physics* V. 1, 224^{b}5, Aristotle appears to deny that a form can be an efficient cause. I suspect that the explanation is that *Physics* V was written before *Physics* II.

16 As he recognizes, one can also ask 'What is it?' of an item of any kind, including not only the species (forms) of primary substances but also qualities, quantities, relations, and so on. (The *locus classicus* is *Topics* I. 9, 103^{b}27–35.) In these latter cases one is asking 'What is it?' of a universal, and the answer sought really is a definition; but where the question is asked of a particular, it is not. For Aristotle constantly adheres to the view that particulars cannot be defined (e.g. *Posterior Analytics* II. 13, 97^{b}26; and most fully *Metaphysics* Z. 15, *passim*).

17 The different classification of causes in *Posterior Analytics* II. 11, 94^{a}20–4, also calls form (i.e. essence) a cause.

I think this is right, but it must be confessed that in *Physics* II. 1–2 he never quite says this explicitly, and the arguments that he presents for saying that matter and form do each contribute to a thing's 'nature' do not make any obvious use of this idea. Again, we cannot rely simply on the point that matter and form are both called 'causes' in the fourfold classification of causes, for as I have observed he does not actually seem to relate them in that discussion to any appropriate 'why'-questions. What, then, do they explain? While one can certainly conjecture aspects of natural behaviour that Aristotle may have taken to be explained by matter and by form (e.g., in the simple case, just by shape), he gives no such examples himself. Nevertheless, I think we can be sure that Aristotle did think of a thing's form as *explaining* something about it.

The most obvious point to fasten on here is that a thing's form is said to be, in many cases, the same as its purpose, and undoubtedly Aristotle did (quite rightly) take purposes to be explanatory. We explain many of the features of a manufactured object by saying what it is *for*, and in chapters 8 and 9 Aristotle insists that the same applies to natural objects too. I shall take up this line of thought in more detail in Section 3. But here what I wish to draw attention to is how this feature of his thought again conflicts with what we have had before.

It seems very sensible to say, of many artefacts (e.g. a tin-opener), that one does not know *what it is* unless one also knows *what it is for*. It also seems quite sensible to say this of several of the parts of animals (e.g. the eyes, the ears), but it becomes progressively more doubtful as the purpose becomes less obvious. (For example, no one knew what the heart was *for* until William Harvey discovered the circulation of the blood in 1616, and what that in turn was *for* was not known until later still. Should we infer that Aristotle, for example, did not know what a heart *is*, or what blood *is*?) One can, of course, pitch the standards of knowledge so high that 'knowing what it (really) is' becomes a very rare attainment, but this surely has little connection with how Aristotle first used the question 'What is it?', and took it to be answered by 'a man' or 'a horse' or 'an oak tree'. Nor is it at all closely related to knowing the appropriate criterion of identity for the thing, or to knowing how the pre-existing matter was changed at the thing's creation. This is really a quite different notion of 'form'. Moreover, most people—including those of Aristotle's day—would deny that the notion has any application to such natural objects as a man, or a horse, or an oak tree; there is nothing that these things are *for*, in the sense which Aristotle seems to intend. (Of course they may be used, by men, for *men's* purposes, but that is not to say that they have a purpose 'of their own'.[18]) I shall explain in the next section why he evidently thought differently.

I add that there is a further way in which form is said to be a 'cause' in this book of the *Physics*, namely as an 'efficient' cause in natural generation. That is, 'man begets man' (or, more generally, 'in the reproduction of living things, what

[18] I here set aside, as quite untypical, the implications of *Politics* I. 8, 1256^{b}15–22.

is produced has the same form as what produces it'). This may seem simply to be a repetition of the earlier thought that a form is a (biological) species, but we shall find in Section 4 that it goes much further than that. Aristotle makes quite detailed use of the notion of form in his theory of (sexual) reproduction and heredity, and this use is not clearly related to any of the ones so far considered.

3. BIOLOGY I: FORM AS GOAL

In *Physics* II. 8 Aristotle argues that we must recognize that nature is a final cause, a 'that for the sake of which' (198^b10–11). His most successful argument, which occupies the bulk of the chapter, is the parallel which he draws between nature and human skill (τέχνη, 199^a8–$^b33$).[19] Human beings make things for a purpose, and Aristotle argues that we will not understand natural objects, in particular living things, unless we recognize that in their case too things happen for a purpose. The argument is applied in two ways, first to the parts of animals and plants. (For example, the front teeth are sharp and the back teeth by comparison flat. Why? Because the front teeth are for biting, where sharpness is useful, and the back teeth for grinding the food, where flatness is useful, 198^a23–9. Compare 199^a27–30 on the purpose of roots and leaves in plants.) But it is applied also to their natural behaviour. (For example, birds make nests for a purpose, and spiders make webs, and ants cooperate with one another, 199^a20–3, 26–7.) It is obvious in the first case that this is not because the animal has *planned* its teeth, or the tree its roots, and Aristotle thinks that the same applies to the second case too: we should not attribute to birds and spiders and ants any planning of their natural activities (199^a20–3). Nevertheless, things in nature are as if some person had planned them, for they do serve purposes, and cannot be explained without mentioning those purposes. As I have said in Essay 4, this seems to me to be a good argument.

I think it is clear that in biology the question 'What is it for?' is very often apposite, and a suitable answer is felt to be explanatory. But the question, as Aristotle sees it, presupposes the existence of a goal, as we may illustrate in this way. What is the stomach for? It is for digesting food. And what is that for? It turns the food into a form that is suitable for nourishing the body. And what is nourishment for? Here we explain (if we can) how without it the body cannot go on living. Well then, what is living for? Here we stop. Living is not *for* any further purpose; it is itself the goal. Generalizing, all the parts of an animal or plant, and its natural activities, are *for* the life that it leads. That is where this line of explanation stops: it is life that is the ultimate goal, and here we of course mean whatever kind of life is typical of whatever specific kind of plant or animal is in question.

[19] I have discussed the argument of this chapter at some length in Essay 4, pp. 49–55.

In the Biological Works Aristotle does not often say explicitly that it is (a certain kind of) life that is the ultimate goal, and there are times when he says something rather different. At *De Generatione Animalium* 717^{a}22–4 he appears to say that 'the' function (ἔργον) of a living thing is to reproduce its kind; at least, he says that this applies to plants, and that it is 'almost' the only function of 'most animals' (cf. 718^{b}9–10, 731^{a}25–^{b}8). But in the more careful statement at *De Anima* II. 415^{a}25–^{b}7, he says rather that this is *one* of the functions of living things, albeit the one that is most natural and most universal.[20] In any case, the 'goal' which is the ultimate explanation of why an animal or plant has the parts that it does is not properly described as the 'function' of that animal or plant, in the same sense as we can assign 'functions' to its parts. They are instruments serving a purpose that is useful to it, but we do not have to suppose that it must also be an instrument, serving a purpose that is useful for something further, for the chain of questions 'And what is *that* for?' does come to an end.[21] I think the best statement of Aristotle's doctrine is at *De Partibus Animalium* I. 5, 645^{b}15–20, which says that just as each of the parts of the body has a function, so too does the *body* as a whole, for it is an instrument that exists for the performance of some complex action (cf. I. 1, 642^{a}12–14). This complex action is evidently living a certain kind of life, for that is closely associated by Aristotle with having a certain kind of soul, and he at once goes on to gloss what he has just said as claiming that the body exists for the sake of the soul. This is a thought I shall pursue later in Section 5. Meanwhile, let us come back to the idea that living is a *goal*.

Ordinarily, to say that something is a goal is to imply that it is *aimed at*, but Aristotle must be denying this implication in the present case. Certainly, he does not think that animals and plants *desire* their life, and for that reason *work out* how to achieve it. Nor does he suppose that there is anything else (e.g. God) who desires their life and plans accordingly. So what entitles us to regard this as a goal? Part of the answer is what we have had already: there evidently are purposes in nature, and purposes presuppose goals. But Aristotle wants to add something else: a goal is something that is *good*. After all, humans adopt certain goals only because they think them good, and though in nature there is nothing that *thinks* things good, still there are things that *are* good, and this is sufficient. So what we are asked to grant is that it is good, *for* the animal or plant concerned, to live the kind of life that is typical of its species (*Physics* II. 7, 198^{b}8–9). That will allow us to view this life as a goal, for the sake of which it grows and develops as it does. Moreover, this is basically how Aristotle's teleological explanations do in practice

[20] The qualification appears to be overlooked by Ackrill (1981: 53), who I think puts far too much weight on this passage. (In any case we cannot take the suggestion seriously if 'function' is to be equated with 'essence', i.e. what a thing is. For example, one cannot explain what an oak tree is by saying that it is for generating further oak trees. That would be blatantly circular.)

[21] In consequence, the well-known claim of *Nicomachean Ethics* I. 7, that man has a 'function' (and the good man is one who performs that function well, 1097^{b}24–33), is at best misleadingly expressed.

work: they show why it is *good* for the animal to have such-and-such a part, and this is taken by him (and by us) as explaining why the part is there.

But now let us ask: what has this to do with the 'form' of that animal? The first response is that Aristotle is committed to invoking form at this point, for he has already assumed (in *Physics* II. 1–2) that the 'nature' of a natural object will be the 'cause' of its natural changes, and that this 'cause' must be either its matter or its form. But the goal is the cause (i.e. final cause) that explains why it naturally grows and develops as it does, and this cannot be ascribed to its matter.[22] It must, therefore, be due to the form. But what has this to do with the other ways in which Aristotle uses the notion of form? Well, there are some connections, but they are not as strong as Aristotle seems to suppose.

Let us first take up the equation between form and essence, i.e. what a thing is. As I have said, in the case of an artefact that has a function, it is plausible to hold that knowing what the thing is for is at least part of knowing what it is. But it is not the whole of it, and one reason is that what we call different artefacts may yet have the same function—e.g. a cut-throat razor and a safety razor and an electric shaver, or again a bicycle and a tricycle. With the parts of living things it is rather the opposite difficulty that is more prominent: what we call the same part may have different functions in different species. Consider, for example, tails. One might very plausibly claim that the function of a cow's tail is to swat flies, but clearly the function of a monkey's tail or a kangaroo's tail is very different. What it is to be a tail is not specially associated with any one function.[23] Again, there are parts of animals which have no useful function, e.g. the appendix in man, and Aristotle too would accept that this may happen. For example, he says of the bile by the liver that it seems probable that it has no function. (It is produced by the operation of the laws of matter as a 'residue', but nature has failed to find a good use for it, *De Partibus Animalium* 677^{a}13–18.) But he evidently would not wish to infer that there is no such thing as what it is to be bile. There are also, as I have said, plenty of examples of parts which do have functions, but unobvious ones, so we can learn to identify the part readily enough without at all knowing its function. I concede, then, that there is *some* connection between knowing what a thing is and knowing what it is for, but it is not identity.

Much the same applies when we turn from what may properly be called 'functions' to what Aristotle regards as the 'goals' of whole animals or plants, namely their ways of living. Things of different species may well exhibit what one is inclined to say is the same way of living (e.g. an oak tree and a beech tree), and things of the same species may live very differently (e.g. a feral cat and a domestic cat). So, to know what it is to be an animal or plant of this or that kind, it is neither necessary nor sufficient to know how the thing lives. Moreover, it surely is a requirement that one knows its 'form' in the quite different sense of its overall

[22] Essay 4 gives some detailed consideration to this claim, which I here simply assume.

[23] As Aristotle recognizes at *PA* IV. 10, 690^{a}1–4.

shape and structure, for that is obviously important for practical purposes of identification. It may well be suggested that *some* knowledge of the thing's lifestyle is needed, for a reason which I have earlier suggested may be relevant to Aristotle's notion of form as what a thing is, namely in order to know what counts as the same animal or plant over time. For example, when a caterpillar emerges from its chrysalis as a butterfly, we do still have the same animal, but when what emerges is an ichneumon fly, we do not. One needs, therefore, to know what kinds of changes are typical of the animal in question, and still preserve the same animal. But this is not to say that we need to know everything about its lifestyle; still less is it to say that we need to know the function of each of its parts, though Aristotle does at one point suggest this (*De Generatione Animalium* v. 1, 778^a16–$^b20$. I shall come shortly to the context in which this point is made; meanwhile, we perhaps should not press it.) I conclude as before that there is *some* connection between essence and way of living, but it is not identity.

A rather different point about form as essence and form as goal is this. Presumably one can know everything that there is to be known about the shape and structure of a particular kind of animal, and what its parts do, and how it lives, without at any point introducing the notion of goodness. After all, a modern biologist would not use that notion, and yet he surely knows the essence of that kind of animal, i.e. he knows what it is. But he would not say that its way of living can be called a 'goal', as if it were something aimed at, on the ground that it is 'good' for the animal concerned to live that way. Knowing what it is, and knowing what is good for it, are quite different ideas, and a modern biologist will not think that the latter has any real application in science. (The evolutionary scheme of explanation does not require the notion of goodness.) Of course, Aristotle's view is different, for he does think that the notion of goodness has a crucial role to play, and his teleological explanations could not function without it. So it is built into his notion of a goal that achievement of a goal is a good thing. But there need be nothing good about what a thing is, and for this reason too goals and essences should not be identified. To put it briefly, what is acknowledged as a goal can, just for that reason, be invoked to explain *why* things happen; but an essence does not explain why things happen, though it may imply *that* they do.[24]

Let us turn more briefly to a different account of what Aristotle calls 'form', namely as what the matter acquires when the thing comes into existence. In one way this seems promising, for indeed what the matter acquires is life. But in another it is awkward, for when a vegetable seed germinates, or an animal is conceived, the matter in question obviously does not at once acquire the lifestyle that Aristotle thinks of as the goal, i.e. the lifestyle typical of the mature adult of the species. One might say, rather, that the relevant new property that comes to be 'in' the matter, in much the same way as a new shape may be said to be 'in' it,

[24] See Essay 4, p. 62, for a fuller statement of this claim.

is a *tendency towards* that goal. (The cause of this tendency we explain in material terms, as the DNA then formed,[25] but it is hardly surprising that Aristotle could not see how such a materialist explanation might be possible.) But is the 'form', then, the (shape and) lifestyle that will eventually develop, if all goes well, or the tendency towards it? For the most part, Aristotle seems to take the former view but I think without really distinguishing it from the latter, though he does so occasionally.[26] Which *should* he say? Well, undoubtedly the goal is the mature state, which is not present from the start, and in some cases is never attained. But if the form is rather what causes the animal to develop towards that goal, then presumably it is present from the start (as, on our view, the DNA is). Or, to take a different line, if the form is the species then again it is present from the start, for an animal begins to be of a particular species—say, a horse—as soon as it begins its existence.

I think one must return a verdict of *non liquet*. There are already too many disparate strands in Aristotle's notion of form. But we have more to come.

4. BIOLOGY II: FORM AS INHERITED

Aristotle's theory of how sexual reproduction takes place in animals is that the male parent supplies the form (conveyed usually by the semen), and the female parent supplies the matter (which he takes to be the menstrual fluid). The theory is developed and argued for in *De Generatione Animalium*, principally in I. 17–23 and II. 1–3. I shall not linger over the arguments that he produces in favour of this theory, for many of them seem to us rather blatantly 'sexist', simply assuming the general inferiority of the female sex. (But he does offer *some* arguments of a more appropriate kind.[27]) Similarly, I shall not linger over his attempts to explain just how form can be conveyed, via semen, to new matter, for these are mainly analogies which seem to us quite unhelpful.[28] Suffice it to say that the overall theory is as I first stated.

[25] At least, in the case of (sexually reproducing) animals life begins when the DNA is formed, though with vegetables the story is rather different, for a seed may lie 'dormant' for a long time.

[26] Usually he says that at conception the matter acquires a new form (from the father), but sometimes he says rather that it acquires a new potentiality (δύναμις), which is presumably the potentiality to grow towards what is properly its form (e.g. *De Generatione Animalium* 729^{b}28, 730^{a}2, 14, 736^{a}27, 739^{a}17, 740^{b}30). Again, he says in the same work that the relevant kind of 'soul'—perceptive, locomotive, etc.—cannot be said to be present until the animal begins to act in the relevant way (736^{a}30–3, ^{b}21–6).

[27] He observes that in several insects that copulate the female inserts a part of her body into the male, and no matter appears to pass from the male to the female (721^{a}11–17, 723^{b}19–25, 729^{b}22–33, 730^{b}25–32). He also thinks he has reason to suppose that in birds and fishes that lay eggs the male semen fertilizes the egg without entering it (729^{b}33–730^{a}28). So, in these cases at least, the male provides no matter, and therefore presumably in all cases. But, according to his standard assumption, if the male contribution is not matter, it must be form.

[28] The semen is 'concocted' from the blood. The blood contains certain 'movements' (κινήσεις), which have developed and now maintain the man's form. The semen contains these too, and

But what exactly is this 'form' which is conveyed? If all that had to be explained were simply that offspring resembled their fathers *in species* ('man begets man') then we could identify 'form' with 'species' and there would be no further problem. We could no doubt add that the mother too must be of the same species, in order to provide matter with the correct 'potentialities', without occasioning any surprise. But the facts of heredity are far more complex than this. Offspring generally resemble their parents much more than they resemble other members of the species, and they may of course resemble their mothers rather more than they resemble their fathers, most conspicuously in sex, but also in other ways.[29] Aristotle tries to show that his theory can after all account for these facts of inheritance in IV. 1–4, but it must be confessed that he is not very successful.

He begins, in chapters 1–2, with the question of the sex of the offspring. Here his explanation is that when all goes well the form provided by the male parent will indeed generate a male offspring, but it may happen that the 'principle' (i.e. cause, ἀρχή) supplied by the male parent fails to 'master' (κρατεῖν) its material—'through lack of heat it cannot concoct it, and lead it to its own special form, but is worsted in this attempt'—and when this happens the material instead takes on the 'opposite' form, the opposite to male being female (766^{a}18–24; cf. 766^{b}15–17). So far, this may sound reasonably plausible. Apparently we have only invoked the fact that male and female are 'opposites', and have not required any input of form from the mother. But this attitude cannot be maintained when we progress to the fuller theory of heredity in chapter 3. Aristotle *tries* to maintain it by the specious reasoning that as father and mother in general may be counted as opposites, so also this particular father may be regarded as opposite to this particular mother (768^{a}6–9). Hence when any particular aspect of the father's form—say, the shape of his nose (768^{b}2)—fails to be reproduced in the offspring, the first expectation is that it will be the corresponding aspect of the mother, counted as 'opposite', that will be reproduced instead.[30] But clearly this is quite illegitimate. This particular mother is no more 'opposite' to this particular father than any human is 'opposite' to any other, and it is (silently) being admitted that the mother may in fact contribute just as much to the *form* of the offspring as the father does. So as a matter of fact the evident data of heredity should have revealed, even to Aristotle, that his theory will not in the end work out.[31] But I set that criticism aside in order to draw out a different implication of the theory.

'communicates' them to the matter provided by the woman. The analogies are designed to show how movements can convey form but not matter, e.g. the form of a bed from the carpenter's mind to the wood he works on (729^{b}9–22, 730^{b}5–32, 734^{b}31–735^{a}5, 740^{b}26–30).

[29] They may also, Aristotle notes, resemble one of their ancestors rather more than they resemble either of their immediate parents (767^{b}3).

[30] Aristotle also offers a theory as to why it may be neither of these, but the nose of the father's (male) ancestors, or of the mother's (female) ancestors. I pass over this.

[31] I also remark that his theory cannot account for something else that he, at least, believed in, namely the 'spontaneous generation' of certain animals. See Essay 4, pp. 60–1.

Suppose that Socrates and Coriscus each beget a child, and—from Aristotle's point of view—things go well. That is, the form conveyed by the semen of each does 'gain the mastery' over the matter supplied by their respective wives, and each has a son. Furthermore, if things go *really* well, then in each case the son bears a striking resemblance to its father, for:

The father is not only male, but a male of a certain sort, e.g. Coriscus or Socrates. Equally, it is not only Coriscus, but also a man . . . and an animal. But [being] a man is nearer to what is peculiar to him than [being] an animal, and the generator is both the individual and the kind to which he belongs, but is more the individual, for that is the substance. (Also, what is generated is indeed a thing of a certain sort, but also a this, and that is the substance.) For that reason, the movements[32] in the semen are derived from all such powers, and even potentially from those of the ancestors, but always more from what is nearer to some individual, such as Coriscus or Socrates. (767^b24–768^a2)

Thus semen from Coriscus does convey the features characteristic of the species man (and of the genus animal), but it also conveys features characteristic of Coriscus in particular, as the semen of Socrates does not. This supposition is necessary to explain heredity. But what the semen conveys is form, and hence the form of Coriscus is not the same as that of Socrates, though indeed both belong to the same species, namely man.

It has been objected to this interpretation[33] that in the crucial chapter 3 of book IV Aristotle never explicitly *says* that everything that is conveyed in the semen derives from the form of the father and contributes to the form of the offspring. That is perfectly true, but not, I think, of any importance. There are equally long passages in the earlier discussion of how animals are generated (especially in II 1–3) where the word 'form' is never used during the detailed description of how the semen acts upon the menstrual fluid. But it is clear that the mechanism Aristotle is invoking here is exactly the same as the mechanism he invokes earlier, and that earlier account is explicitly an account of how form is transmitted. Besides, Aristotle evidently still retains the view that the male parent contributes no matter, and nowhere suggests that part of what he contributes counts as neither matter nor form.[34] Moreover, he *does* refer explicitly to form on either side of IV. 3. Discussing the formation of female offspring in IV. 1 he says that it happens when the male principle cannot concoct the matter into 'its own form' (τὸ ἴδιον εἶδος, 766^a20), and adds that this results in something 'differing greatly in form' (πολὺ τῷ εἴδει διαφέρει, 766^a25). (He at once goes on to compare the feminine appearance of the eunuch.) So it is clear that he is here counting the male and the female of the same species as differing in form. Then

32 For these 'movements' see n. 28 above.

33 The interpretation was brought into prominence by Balme (1980). It is disputed by Witt (1985).

34 Witt (1985) claims that where Aristotle earlier anticipates his theory of inheritance at I. 19, 726^b13–18, he distinguishes *two* functions for semen to perform, but I see no such distinction in this text. (Note that it occurs *before* we have been told that what the male contributes is not matter but form.)

when he turns to the more general theory of inherited resemblance in IV. 3 he begins with the statement that this is due to 'the same causes' (767ᵃ36), and at the end of the chapter he passes without a break to the discussion in IV. 4 of various kinds of deformed offspring, still offering exactly the same kind of explanation. And in the course of this discussion he remarks that such deformities are in a way not 'contrary to nature' since they occur when 'the nature in accordance with form fails to master the nature in accordance with matter' (770ᵇ16–17). The same theory is being applied all through.

There thus seems to me to be no temptation to say that the theory of inheritance is no longer being conceived as a theory of inherited *form*. No doubt the most important aspect of this form, that the male contributes to the matter supplied by the female, is life (or soul), and an appropriate 'principle' to ensure that the newly living matter will grow and develop in a suitable way. But to accommodate the detailed facts of heredity we must construe this 'form' as including also the shape and general appearance of the new animal, and indeed all evidently inheritable characteristics. So form, though still perhaps considered as in principle universal, must now be construed as a very specific universal indeed, and almost any two actual animals will be found to exhibit *some* difference of form, save perhaps for identical twins.

If 'form' is thus to be distinguished from 'species', should we also separate it from 'essence'? This at any rate is the implication of the opening of book V of the *De Generatione Animalium*, which tells us that the essence (λόγος τῆς οὐσίας) contains all and only those features of the animal that serve some purpose, together with the purposes served (778ᵇ12–14; cf. 778ᵃ34–5). The book then goes on to discuss various features of animals that are not part of the essence, for example the colour of the eyes, or of the hair, the curliness or straightness of the hair, and so on. During his discussion of these features Aristotle never does in fact remark that they may be inherited, though his account would allow for this, and he apparently makes provision for it in his opening statement ('The cause should be traced back to the matter *and* to the source which initiated the movement', 778ᵇ1–2). In any case, it is obvious that many such features *are* inherited, and so will be part of the form but not part of the essence, according to the present criteria. But it is noticeable that Aristotle himself never does explicitly draw this (or any other) distinction between form and essence, and it is quite possible that he never became aware of this consequence of his theory of inheritance, or that if he did he very soon forgot it.

In any case, the main moral of this discussion is obvious: the form 'man' is here no longer being regarded as 'indivisible', for now it does subdivide into male and female, and apparently also into black men and white men, men with snub noses and men with hooked noses, and in countless other ways invoking other features that can be inherited. This is a very notable departure from what we had initially in the Logical Works. But there is yet to come what seems to be an even more drastic revision.

5. THE *DE ANIMA*: FORM AS SOUL

In many places where Aristotle speaks of 'soul' (ψυχή) we would more naturally speak simply of life, and at least it is clear that soul and life always go together for him. Thus living and non-living things are contrasted as those that do and do not have soul (τὰ ἔμψυχα, τὰ ἄψυχα), and in this contrast plants certainly count as living, and so also do the parts of living bodies. For this reason, we have already heard quite a lot about the soul during Aristotle's account of how animals are generated, for when the semen imparts a 'form' to the matter provided by the female what it principally does is to give *life* to that matter. But in this section I shall concentrate upon Aristotle's main account of the soul in the *De Anima.*

Soul in general is correlated with life in general, but Aristotle also insists upon our recognizing different kinds of soul, or life. He gives a brief catalogue in *De Anima* II. 2. Lowest in the scale is what he calls 'the nutritive soul' (ἡ θρεπτικὴ ψυχή), which is primarily correlated with being nourished by food. But he also associates it with growth and diminution (413^a25, 416^a25–7) and with reproduction (415^a23, 416^a19). Next comes 'the perceptive soul' (ἡ αἰσθητικὴ ψυχή), which is also associated with imagination (φαντασία) and with desire (ὄρεξις), on the ground that where there is perception there is also pleasure and pain, and where these are there is bound to be desire (413^b22–4). After this comes 'the locomotive soul' (ἡ κινητικὴ κατὰ τόπον), which is found in animals that are capable of moving about from place to place, and finally 'the thinking soul' (ἡ διανοητική). If we may confine attention to mortal and perishable things, these various kinds of soul are hierarchically arranged. Thus *every* living thing has the nutritive soul, including plants, whereas it is definitive of animals that they also have perception, at least in its most primitive form, namely touch. (This is also taken to include taste, 414^b6–14, 422^a8.) Some animals also have higher forms of perception, of course, and again some have the power of locomotion.[35] But no animal except man has the power to think, and he has *all* the kinds of soul here distinguished.

It would seem that the way Aristotle is thinking is something like this. All things that are alive exhibit life by taking in food, whereby they first grow and then maintain themselves, and also by reproducing. Animals also exhibit life by perceiving their environment (and by reacting to it appropriately), and some of them also by moving about. One kind of animal, namely man, manifests life in all these ways, and by thinking as well. To be alive just is to do these things. To say, then, that a thing has this or that kind of soul just *is* to say that it has a life of the appropriate kind, and the 'soul' in question simply *is* that way of life. It is not, in our sense, the *cause* of that life, something which *brings it about.* In Ryle's

[35] The precise hierarchical ordering between the various forms of perception above touch, and between these and locomotion, is left unclear. But 434^b25–7 tells us that any animal that can move about must have some sense by which it can perceive things at a distance.

famous metaphor, the soul is not 'the ghost in the machine', that *makes* the machine work as it does, but is rather the working of the 'machine' itself.

This, at any rate, seems to be a fair inference from Aristotle's explicit definition of the soul in II. 1. He begins by reminding us that a substance is a compound of a certain form in a certain matter, where the matter is a potentiality (*δύναμις*) and the form an actuality (*ἐντελέχεια*) (412^{a}6–10). Then he at once proceeds to say that, in the case of a living thing, the body is its matter and the soul its form. So 'the soul is the essence (*οὐσία*), as form, of a natural body that potentially has life. The essence is actuality, so the soul is the actuality of a body of this kind' (412^{a}19–22). Two points may at once be made in elucidation. First, it is obvious that Aristotle does not mean that the soul is the essence of a body that has life merely potentially, and not actually. He means rather that bodies that are actually alive have soul as their form, essence, or actuality (412^{b}25–6), and he has put in the word 'potentially'—somewhat misleadingly—only because he so constantly associates matter and potentiality. Secondly, he goes on to explain that the soul is 'the first actuality' rather than 'the second actuality', meaning by this that the soul is the disposition to act in ways that manifest life, and not the actual acting in those ways. For example, an animal has life, and therefore soul, even when it is asleep (412^{a}22–8).

So it appears that to have soul is to have life, a kind of soul is a kind of life, and the soul itself simply is that life. That is why we need not ask how soul and body are one, for it is simply in the way that form and matter are one, as when some wax has a certain shape (412^{b}6–9). But a better comparison is where the form of a thing is not just its shape but its function, as with an axe or an eye. For the form or essence of such a thing, what it is to be an axe or an eye, is to have the capacity to perform the relevant function (i.e. chopping or seeing, respectively). The difference is that an axe is merely an artefact, and does not chop by itself, but needs an agent to wield it. In this respect the comparison with the eye is better, but still not entirely appropriate, for an eye can only be an eye (properly speaking)—i.e. it only has its capacity for seeing—when it is part of a larger whole, the whole animal. 'But if the eye were an animal, sight would be its soul; for this is the essence of an eye, as given by its definition' (412^{b}10–22).

It would seem to follow that just as there cannot be chopping without an axe, or seeing without an eye, so there cannot be life, i.e. soul, without a living body. But here we are brought up short. Aristotle is willing to draw this inference for the nutritive soul, and for the perceptive soul, but he is *not* willing to draw it for the thinking soul, and this appears to herald a discrepant theme. I cannot here explore the implications of this point, for the topic is certainly not straightforward, and many problems arise.[36] I therefore set aside the allegedly special nature

[36] So far as the *present* passage is concerned, one may take it that his thought is just that there is a thinking soul which has no body, namely God. (The same interpretation may be offered for some other passages which suggest that the thinking soul, or intellect (*νοῦς*), is separable from the body, e.g. 413^{b}24–7.) But there are also indications that Aristotle thought that a *human* thinking soul

of the thinking soul. But in the same concluding passage in which Aristotle introduces this hesitation, he also introduces another, which we cannot ignore. For the chapter ends:

> That the soul is not separable from the body—or at least certain parts of it, if it has parts—is clear. For the actuality of some parts is the actuality of the parts [of the body] themselves. Nevertheless, there is nothing to prevent some parts [being separable], since they are the actuality of no body. Moreover, it is not clear whether the soul is the actuality of the body in the same way[37] as is the sailor in the ship. (413^{a}3–9)

But, one protests, surely this last point is already quite abundantly clear. If there is any sense in which the sailor in his ship may be said to be 'the actuality' of the ship, it is not the sense in which 'actuality' has been used here. The form, essence, or actuality of the ship is perhaps its shape and structure, or perhaps better its disposition or ability to 'behave' in ways that fulfil its function as a ship (a disposition or ability that it retains even when 'asleep', and empty of any sailor). Even if Aristotle is thinking of the 'actuality' of the ship when actually sailing, this is surely its *sailing*, and not the sailor within it, who is perhaps the efficient cause of the sailing, but evidently not the sailing itself. The image of the sailor very strongly suggests that the soul is the efficient cause of those activities by which the body manifests its life, and not either the activities themselves or the disposition to perform them. It seems amazing that Aristotle apparently does *not* think that what he has said so far has ruled out this image.

If this brief passage canvassing the sailor-image had stood alone, then we could no doubt dismiss it as a temporary aberration from Aristotle's true doctrine, and of no significance. But it is not alone. When we turn to what Aristotle says in II. 4 about the various ways in which the soul is a cause, it appears to be confirmed.

The soul, he tells us, is a cause in three of the ways distinguished, for it is that from which motion starts, that for the sake of which, and the cause as essence (οὐσία). It is a cause as essence because the essence of a thing is 'the cause of its being', and for living things to be is to be living, and the soul is the cause of this. (Besides, we have defined the soul as actuality, and actuality is the definition of what is potentially.) It is also a cause as that for the sake of which, since nature acts for the sake of something, i.e. for some goal, and in animals the goal in accordance with [their?] nature is the soul; for all natural bodies[38] are

could exist without a body, principally at *De Generatione Animalium* II. 3, 736^{b}21–9 (only the intellect can enter the body 'from outside'); *De Anima* I. 4, 408^{b}18–29 (the intellect cannot be affected—is ἀπαθές—by bodily changes due to old age); *De Anima* III. 5 (a desperately obscure account of what Aristotle calls 'the active intellect'). The interpretation of all these passages is controversial, and especially the last.

[37] Ross emends the text to read 'in this way, or as the sailor in the ship'. This does not affect the substantive point.

[38] It seems very improbable that Aristotle really means what our text apparently implies, namely that *all* natural bodies have souls. It is more likely that this is just a slip, and he means to generalize only over the natural bodies of living things. (Ross emends the text accordingly, but I do not find the emendation very convincing.)

instruments of the soul; and the same applies to plants no less than to animals. Finally, the soul is that from which motion from place to place starts, and perception and growth are also 'in accordance with soul' (415^{b}8–27).

First it is worth noting a rather suspicious feature of the argument to show that the soul is a formal cause, i.e. a cause as essence. One expects him to say: 'The essence of anything is what it is for that thing to be; for living things, to be is to be living; and their soul is their life'. But what he actually says is: 'The essence of anything is the cause (*αἴτιον*) of its being; for living things, to be is to be living; and the soul is the cause and principle (*αἴτιον καὶ ἀρχή*) of this' (415^{b}12–14). Now on the assumption that 'cause' in this sentence really does mean 'cause as essence'—e.g. in the way that health is the cause as essence of being healthy—there is no fallacy. But one cannot help noticing how here, and indeed everywhere else, Aristotle shies away from the direct statement that soul *is* life, which would indeed have been rather shocking to an ordinary Greek audience. It seems to me that his definition of soul as 'actuality' does commit him to saying this, or at least something very similar, but he never does. He prefers a circumlocution, as here he says that the soul is the 'cause' of life, which would be more acceptable to an ordinary (Greek) audience just because they would *not* construe 'cause' in this statement to mean 'cause as essence'.

But it is much more significant that here Aristotle very explicitly says that an animal's (or plant's) soul is the *efficient* cause of its various activities. Admittedly, part of what he offers by way of argument is extremely weak; to say merely that growth occurs 'in accordance with soul' is hardly to substantiate the claim that the nutritive soul is the *efficient* cause of growth, rather than simply its final cause.[39] (No doubt we can say that the nutritive soul is an efficient cause in reproduction, in that the soul in *one* plant or animal is 'the starting point' of a process which results in there being a nutritive soul in *another* plant or animal, but it is obvious that that is not what Aristotle means here.) But when he claims that the soul is the efficient cause of an animal's movements, it seems clear what he must mean. What is needed to initiate an animal's movement is a desire, and this in turn may be aroused by a perception. Then if we suppose (as would be a

[39] There are clear statements elsewhere that the nutritive soul counts as the *efficient* cause of growth (e.g. *PA* I. 1, 641^{a}26–7, ^{b}4–10; *GA* II. 4, 740^{b}25–741^{a}3), but no clear and literal explanation of quite how it fulfils this role. It is tempting (to us) to suppose that a final cause can only *explain* why things happen if it can also be seen either as being or as generating an associated efficient cause which is what really makes those things happen. For example, the goal of an action explains that action only because it is really the *desire* for that goal that *begins* the action, and the thought is that final causes in nature must be seen as operating in a similar way. (For this thought see e.g. Irwin 1988, ch. 5; Furley 1996; Lennox 2001*b*, ch. 8.) I doubt whether Aristotle himself would have agreed with this as a completely general claim, but one can see from his detailed work in biology why he should be tempted to it in this case. For to explain why animals grow and develop as they do he frequently invokes a personified 'Nature' acting in their best interests, and this 'Nature' is clearly being thought of as a craftsman acting as efficient cause. But it is officially identified with *the nature of* the animal in question, which in turn is taken to be both the essence and the final cause of that animal. (See Essay 4, pp. 61–3.)

quite ordinary Greek idea) that it is the soul that perceives, and the soul that desires, it will seem natural to say that it is the soul that initiates the movement. But again this embodies a conception of the soul which is quite at odds with Aristotle's definition of it as the form, essence, or actuality of the living body, for it is being viewed rather as some special *part* of the man. Evidently, it is the *man* who desires. If we wish to say that it is, more particularly, his soul that does the desiring then this is as if we had begun with the thought that the man was in contact with the floor and corrected this to say 'more accurately' that the soles of his feet were. But the soul as the 'actuality' of the man clearly cannot be equated with some special *part* of him. That is, this way of talking is one that again suits the image of the sailor in his ship, and is quite inconsistent with the view that the relationship of soul to body is the same as that of shape to matter, or function to implement.

Now I do not think that there is anywhere else where Aristotle explicitly says that the soul of an animal, which is its form, essence, and goal, is *also* the *efficient* cause of that animal's behaviour. But he does very frequently use the idioms that go with this view of the soul, as being like the sailor in the ship. Thus he constantly speaks of the soul as doing various things, e.g. perceiving, discriminating, desiring, and of course thinking. Earlier in *De Anima* I. 4 he has cautioned us that this way of speaking is not strictly correct, and that it is better to say that it is the man who does these things, rather than his soul, but it can hardly be said that he conforms to this recommendation in his own practice. Moreover, when one looks in detail at this passage ($408^{b}1$–18), one finds that it simply reinforces, and does not in any way retract, the idea that the soul is a special part of the man. For his claim is that although we might ordinarily say that the soul 'is moved' by such things as joy, fear, anger, perception, and thought, it is better to say that it is the man and not the soul that is thus 'moved'. This, he goes on, is because the man is moved *by* his soul, for example in anger the soul causes the heart to move. Or again, in perception the relevant movement is one that reaches to the soul, and in memory it is one that reaches from it. So it is not the soul itself that 'is moved', but rather the body, and thereby the combination of soul and body.[40] Here it is quite clear that the soul is being conceived as a special part of a man, and one which causes changes in the body. It suits the image of the sailor in his ship, for he moves it and it does not move him (save 'incidentally', as Aristotle would say, in so far as it carries him from one place to another).[41]

What should we conclude? Either we can say that Aristotle's true view is that the soul is indeed the 'actuality' of the living body, that his stray mention of the sailor in the ship is of no importance, that his calling the soul an efficient cause is

[40] This (quite unreasonable) claim that the soul is 'not moved' leads into the special claim about the separability of the intellect noted earlier (n. 36).

[41] In the *Parva Naturalia* (but not in the *De Anima*) Aristotle specifies the location of the soul as the heart (and not, as we would think, the brain). I think it is fair to say that throughout the *Parva Naturalia* Aristotle views the soul as analogous to the sailor in his ship.

simply a mistake, and that his continued use of traditional terminology means nothing. Or we can say that Aristotle did really think of the soul as like the sailor in his ship, which certainly is the more ordinary conception of the soul (and the one that Aristotle inherited from Plato), and although his official definition of the soul suggests something very different he has failed to appreciate this fact. If we look simply at the quantity of evidence on each side, then I think it is fair to say that in the works which are explicitly presented as treatments of the soul, i.e. the *De Anima* and the *Parva Naturalia*, we have only the official definition in *De Anima* II. 1 on the one side, perhaps with II. 2–3 in support, and almost everywhere else remarks which favour the other.[42] If I may offer a conjecture, I think that it is when Aristotle is bearing in mind his own claim that the notion of a soul applies to *all* living things (including plants) that he thinks in terms of the official definition, but when he is thinking of the soul as a centre of consciousness—as he often does—he reverts to the more ordinary view. I leave that simply as a conjecture, but in any case the truth seems to be that he thinks in both ways.

The consequences for the notion of form are remarkable. The official definition of *De Anima* II. 1 claims that the form of a living body is its soul. But on the ordinary way of thinking about souls they are particulars and not universals. Socrates' soul is something that is located within Socrates (at any rate, while he lives on earth), and is not the kind of thing that could simultaneously be located in some other man. If we are to construe the soul of an oak tree in the same way, then it too is a particular entity, located in that oak tree, a 'part' of that tree which functions as the efficient cause of its growth, of its shedding leaves in winter, of its resistance to frost, and so on. It cannot *also* be a 'part' of any other oak tree, and two trees may be similar in soul, but they cannot literally have the same soul. But the soul of a thing is its form, so forms have now become particulars. Aristotle never says just this in so many words, but there are a few places where it seems to be very clearly implied by what he does say. One such is a passage at the end of *De Anima* I, where he says, of two living things which originate by the division into two of what was one living thing, that each of them 'has the same soul in form (εἴδει), if not in number (ἀριθμῷ)' (412^{a}20–1). But if he can assume without explanation that his familiar contrast between being the same in form and being the same in number can be applied to souls, he surely must be taking those souls to be particulars.[43]

[42] Ross says, in his edition of the *De Anima* (1961): 'the conception of the soul as the entelechy of the living body is . . . the central conception of the *De Anima*' (p. 10), and on this ground he contrasts it with the *Parva Naturalia* and infers that its composition is later. But the truth is that this conception of the soul is absent from almost all of the *De Anima*, save II. 1–3 (It is anticipated in I. 1, 402^{a}26, and recalled in II. 4, 415^{b}15, but that is all.)

[43] The context is long and involved, and not important for my point, so I do not give it. But when it is taken into account one can see that he means that these souls are *exactly* the same in form, since each is a *complete* replica of the one original, but still they differ from one another in number.

Yet Aristotle does not abandon his previous ways of thinking about form. He *still* thinks of form as a species, as an essence (given by a universal definition), as what is acquired or lost in any kind of change whatever, as a function and as a goal (i.e. a way of living towards which an animal or plant will naturally develop, if nothing hinders it). A particular soul, which has its own perceptions, desires, and thoughts, cannot fulfil any of these roles. Nor can it be what is passed from father to son at conception, nor what stands to a living body in the same relation as shape to the wax, or sight to the eye. Yet Aristotle is not willing to abandon *any* of these ways of thinking of forms.

That is why the welter of considerations that are strung together to make up what I take to be his final discussion of form, in books *Z* and *H* of the *Metaphysics*, is so incomprehensible. *All* of these threads reappear in that discussion, and *none* is directly repudiated. There are also yet more, apparently distinguishable, roles of form that are there added. For example, in *Z*. 17 it is described as 'cause of being' and 'cause of unity'. The first of these leads Aristotle to stress again that form is actuality rather than potentiality (especially in *H*. 2), and to try to use this entirely specious contrast to provide serious explanations (especially in *H*. 6). The second connects with his thought that *only* living things have the kind of unity that is required (*Z*. 16, 1040^b5–16), which must lead to the conclusion that the only real forms are souls. This simply disqualifies all of the more straightforward examples that he so constantly relies upon to illustrate what he means.

I add that in this discussion I have had to set aside (as too controversial) the special status that Aristotle assigns to 'the thinking soul', and his obscure hints that this in humans might be immortal, and so might exist without any material body. In any case, we can be sure from book *Λ* of the *Metaphysics* that in his view God is an immortal thinking soul, and does exist without any material body. If the notion of form has to encompass these ideas too, so that there are also forms which are not the form of any matter, then here is another role for what Aristotle calls 'form', which is surely quite distinct from all the others.[44]

If we include the *Metaphysics*, it is absolutely clear that *nothing* could satisfy *all* the many claims that Aristotle makes about forms. But the point that I have tried to bring out in this review is that the same verdict applies even if the *Metaphysics* is excluded. In his many enquiries, Aristotle faces up to a great variety of different problems. He tries to make use of the notion of form in treating almost all of those that fall under the general heading 'problems concerned with nature', but the roles that he assigns to it in the different cases really are very different from one another. Nothing could fulfil them all. For example, we might say (from our

[44] Please note the 'if'. The first and unmoved mover which is Aristotle's God is certainly described in book *Λ* of the *Metaphysics* as 'actuality' (1071^b12–22), and elsewhere Aristotle frequently equates 'actuality' and 'form'. But it is not really clear whether he wished that equation to apply in this particular case.

own perspective) that the biological roles that Aristotle assigns to form may to some extent be regarded as fulfilled by our (materialist) view of DNA: it explains the mechanism of heredity; it governs growth and development; it settles (in so far as anything does) the species of a thing. But of course this applies only to living things, and there is no analogy to the function of a tool or implement, to the various goals that people pursue, or to that 'part' of us that perceives, desires, thinks, and so on. Still less is there a *contrast* between DNA and what everyone recognizes as 'matter', but this supposed contrast is the one feature of form that remains constant throughout Aristotle's many uses of this notion.

There is surely nothing that could play all of the many and differing roles that he assigned to it.

6

Aristotle on the Eleatics in *Physics* I. 2–3

In the first half of chapter 2 of *Physics* I Aristotle explains why a work on physics should pay no attention to the thesis of Parmenides and Melissus that 'what is is one and unchangeable' (184^{b}25–6). But he nevertheless goes on to pay it some attention. The discussions that follow in chapter 2 offer no difficulty in interpretation or in evaluation, so I treat them very briefly in my Section 1. But chapter 3 begins with a criticism of Melissus (186^{a}10–22) that is certainly puzzling at first reading, so I treat this in Section 2. I argue that in this case Aristotle's reasoning can be elucidated in an entirely satisfying way. Then at 186^{a}22 ff. we find a criticism of Parmenides that makes use of the obscure phrase *τὸ ὅπερ ὄν*, and this certainly continues until at least ^{b}12. I treat of this in Section 3, and I argue that we can find good sense in his discussion, even if at times he does not say quite what he should have done. Finally, the overall structure of the discussion from there to the end of the chapter is obscure, and it is not altogether clear what Aristotle means to be attacking. I discuss this in Section 4, but without reaching any very useful conclusion.

1. THE OPENING OBJECTIONS

At 185^{a}20–^{b}5 Aristotle asks how the Eleatic doctrine that 'all things are one' is to be understood, bearing in mind his own doctrine of categories. The right answer is fairly clearly that the doctrine is that there is only one *substance*, though no doubt there are also many attributes which it has, and many which it does not have. But Aristotle does not tell us that this is the answer required, and seems rather to suggest that the Eleatics would have no answer to his question.

At 185^{b}5–25 he goes on to ask how the word 'one' is to be understood in this doctrine, offering as alternatives that 'one' may mean 'one by being continuous' or 'one by being indivisible' (e.g. as points are) or 'one in definition'. Now in fact both Parmenides (fr. 8, 22–5) and Melissus (fr. 10) do say that their one thing is indivisible, but they evidently do not mean to view it as a point. On the contrary, Parmenides at once infers that it must be continuous (ibid), and presumably Melissus would not dissent from this. So, of the alternatives that Aristotle offers, they would do best to accept the first. Aristotle objects to this (^{b}9–11) that what is continuous is infinitely divisible, and so must have parts, which therefore implies that there is more than one thing. His thought is evidently that each of the many parts will exist. But here Parmenides or Melissus might well have replied to him in his own coin.[1] For they could say that so long as the whole is all together, and the various parts are not separated, then only the whole exists and the parts do not. We normally think of the parts as existing because they exist *potentially*, which is to say that they *could* be separated, and so *could* exist. (This is Aristotle's own doctrine. It is briefly indicated here at 186ᵃ1–3, and developed more fully elsewhere, e.g. *Metaphysics Z*. 13, 1039ᵃ3–14.) But in the present case, they could say, the parts cannot be separated, and so do not have even a potential existence. For to separate them one would need to divide what is from what is, which could only be done by inserting what is not, as a divider. But since there is no such thing as what is not, this is an impossibility (Parmenides, fr. 4; Melissus, frs. 7(7) and 10). Here again, then, Aristotle asks a fair question, but he fails to consider with any sympathy how his opponents might answer it.

At 185ᵇ25–186ᵃ3 the final paragraph of the chapter considers the manoeuvres of some people later than Parmenides and Melissus who made a fuss over the problem of how one thing could also be many. Aristotle remarks that this arises in two forms, (*a*) how one thing can have many attributes, and (*b*) how one whole can have many parts. (The avoiding manoeuvres that he mentions concern (*a*) rather than (*b*).) In each case he seems to reply that we have what is in *one* way one, and in *another* way many. At any rate, this is clearly his answer to (*b*), namely that what is *actually* one (whole) may also be *potentially* many (parts). I think he means to suggest the same approach for (*a*) also, namely that the pale thing and the artistic thing are one *accidentally* but two in *definition*,[2] but our text fails to spell out just how they are one.

2. THE ATTACK ON MELISSUS

At 186ᵃ4–18 Aristotle raises some objections specific to Melissus' line of argument. The first and main objection is this:

It is obvious that Melissus reasons fallaciously. For, assuming that everything that has come into being has a starting point (ἀρχή), he thinks he has also assumed that what has not come into being does not have one. Then this too is absurd, that there should be a starting point of everything—of the thing, not of the time—and that this applies not [only] to simple coming to be but also to alteration, as if change does not take place all over. (186ᵃ10–16)[3]

[1] In fact Melissus' answer appears to be different: the one is not a body, and only bodies have parts (fr. 9).

[2] See e.g. *Topics* 103ᵃ33–9, 133ᵇ17–36; *Sophistici Elenchi* 179ᵃ32–ᵇ4. On the implications of this notion of 'accidental identity' see e.g. Matthews (1982).

[3] This translation does not differ in any serious way from those proposed by most others (e.g. Ross 1936; Hardie and Gaye 1930; Charlton 1970). It does differ seriously from that given by Gershenson and Greenberg (1961), but I think it is clear that their version is untenable.

There are a number of obstacles to understanding this passage. Let us begin at the beginning: why should Aristotle make such a fuss about the illicit conversion that he puts first? Even if Melissus' premiss does not imply that what has not come into being has no starting point (or beginning), still the proposition appears to be true anyway. So would not Melissus be entitled just to take it as a further premiss?

The first thing to note is that just this accusation of illicit conversion is Aristotle's *standard* objection to Melissus; he repeats it three times during his discussion of fallacies in the *Sophistici Elenchi* (167^b12–20, 168^b35–40, 181^a27–30). In the first of these passages Aristotle also tells us that the subject to which Melissus applies this reasoning is the universe (τὸ ἅπαν): he has argued that the universe has not come into being (since nothing can come from nothing), and hence infers that it does not have a starting point (ἀρχή). But in all three passages we also find a further thought credited to Melissus, on which Aristotle offers no adverse comment, namely that since it has no starting point it is infinite (ἄπειρον, 167^b26, 181^a29; not πεπερασμένον, 168^b40). Now this word 'infinite' can of course be used of what is infinite *in time*, and a good reason for taking it in that way here is that the claim that what has no starting point in time must already be temporally infinite in one way (i.e. backwards) would seem to be undeniable. Moreover, it would be a standard Greek assumption that this would also carry with it being temporally infinite the other way (i.e. forwards). But on the other hand the word is more usually used of what is spatially infinite (as e.g. at 185^b16–19 here in the *Physics*), and besides it is spatial infinity that is at issue between Aristotle and Melissus. For both agree that the universe is temporally infinite (in both directions), but Aristotle thinks that it is spatially finite while Melissus thinks that it is spatially infinite. So one would certainly *expect* Aristotle to attack Melissus on his deduction of spatial infinity, and not to fuss overmuch about any faults there might be in his deduction of temporal infinity. Moreover, in our present passage from the *Physics* (though not in the *Sophistici Elenchi*) there is a strong indication that it is not time that is in question, since it is explicitly said that the 'starting points' (ἀρχαί) that we are talking of are starting points 'of the thing, *not* of the time'. Of course, this still leaves much unexplained. It is not at all clear what a 'starting point of the thing' is supposed to be, nor why what lacks such a starting point would have to be spatially infinite. To answer the first of these questions, I turn now to the second sentence of our passage.

Aristotle says: 'This too is absurd, that there should be a starting point of everything.' But of course Melissus did not hold that there was a starting point of everything, so Aristotle cannot have said quite what he meant. But the remedy is obvious: with 'everything' we are to understand the qualification 'that has come into being' from the previous sentence. Then in the first sentence Aristotle is objecting that Melissus' premiss does not imply his conclusion, and in the second sentence he is saying that the premiss is absurd anyway.

I believe[4] that he thinks it is absurd even for simple coming to be (i.e. for coming into existence), which is the primary application that Melissus desires. But he thinks it is more obviously absurd for alteration, since there is such a thing as 'change all over' (*ἀθρόα μεταβολή*). There is just one other place in Aristotle's writings where this strange phrase is used, namely at *Physics* VIII. 3, 253^{b}25, and consideration of the surrounding context (^{b}14–31) makes it quite clear what is intended. Aristotle is saying that with some alterations—and he gives freezing as an example—what is altered is divisible into parts, but we need not suppose that the alteration is itself divisible in a similar way. That is, we need not suppose that it begins with one part and spreads continuously from there to the others; it may occur 'all over', i.e. it may affect all of the water's surface simultaneously.[5] In such a case, then, there would be no starting point for this change, i.e. there would be no starting point 'of the thing', though no doubt there would be a starting point 'of the time'. The same applies to coming into existence. There is no need to suppose that there must be one part of the thing that comes into existence first, and then another, and then another, and so on. Indeed, this cannot be so if the thing in question has no parts (e.g. a point[6]), and it need not be so even with things that do have parts.[7] A starting point of a thing, then, is a part (or point?) of the thing that comes into existence before its other parts. Aristotle says that we need not suppose that what has come into being has such a starting point. But he *also* seems to grant to Melissus that what does not have such a starting point must be infinite. Why should he do that?

Well, the answer appears to be that Aristotle has only given us half of the argument of Melissus that he has in mind. For Melissus had claimed:

> If it had come into being, it would have had a starting point (ἀρχή), for it would at some time have started to come to be, and it would have had a finishing point (τελευτή), for it would at some time have finished coming to be. But since it neither started nor finished, it always was and always will be, and has neither starting point nor finishing point.[8]

The suggestion is, then, that Melissus had argued: if it had come into being, this process of coming into being would have taken some finite time. So there would have been a first part of it to come into being, *and a last*, and the *latter* implies that it has only finitely many parts, and so is finite in extent. Thus we have an argument purporting to show that what has come into being must be finite in extent. Aristotle thinks that this argument is doubtful; it is not clear that if the universe had come into being it would have to have had such first and last parts.

[4] I have to admit that the text is ambiguous on this point, and that is why my translation has the word 'only' in square brackets.

[5] Incidentally, this *denies* what is claimed in the earlier book VI, particularly chapter 4. See my (1991: 200), this volume, p. 175–6.

[6] *De Caelo* 280^{b}25–8; *Metaphysics* 1002^{a}28–^{b}11, 1044^{b}21–2; *Nicomachean Ethics* 1174^{b}10–13.

[7] Contrary to what is claimed in VI. 6, 237^{b}9–22.

[8] Fr. 2, lines 3–7. But my translation assumes that Ross is right (1936: 472) in twice correcting DK's γενόμενον to γινόμενον.

(But note, incidentally, that every creation myth does seem to conform to the assumption.) His main objection, however, is much more powerful. For suppose we grant the conclusion that what has come into being must be finite in extent, and suppose we grant also (for independent reasons) that the universe has not come into being. Still, it is just a fallacy to infer from this that the universe is infinite in extent. And this fallacy is crucial to the major difference between Aristotle and Melissus on the infinity of the universe. But where he has puzzled us, his readers, is that he has levelled the charge of illicit conversion, not at the conclusion of Melissus' argument, but just at its first step, so that we do not easily see the significance of the point that he is making.

For a deeper evaluation of Aristotle's criticism of Melissus, we should have to consider whether Melissus did really argue as I have been suggesting. I think it is clear that he could be *taken* as arguing in this way, and that it is not at all surprising that Aristotle *did* so take him, but one cannot be more definite than this without going further into the evidence on Melissus than would be proper here. So I leave that issue unresolved. Meanwhile, I bring this section to a close with some very brief remarks on Aristotle's remaining objections to Melissus.

At 186ᵃ16–18 he makes a fair point about motion. Melissus had argued that there could be no motion, because motion requires empty space, for the moving thing to move into (fr. 7(7)). Aristotle replies that rotations require no such empty space, nor does any other motion in which each moving thing is moving into a space being vacated by another moving thing, as e.g. with currents in water. Melissus has no answer to this. Then at 186ᵃ18 Aristotle merely asks why Melissus had denied alteration to what is, but since he fails to provide or discuss the reasons that Melissus did give (e.g. in fr. 7(2)), there is no argument here to evaluate. Finally at 186ᵃ19–22, apparently breaking off any dispute with what Melissus himself did say or might have said, Aristotle remarks that since man and horse differ in form we cannot even say that everything is 'one in form'. This appears to be a kind of footnote, relating to something that was not said earlier at 185ᵇ5–9, where one might have expected a mention of things that are one 'in form'.

3. THE MAIN ARGUMENT

At 186ᵃ22–ᵇ12 Aristotle presents his main argument against Parmenides' claim that there is only one thing. It breaks naturally into three sections, ᵃ22–32, ᵃ32–ᵇ4, ᵇ4–12.

186ᵃ22–32. The first section opens with the point that Parmenides assumed a false premiss when he assumed that 'what is' (*τὸ ὄν*) had only one meaning (ᵃ24–5). Now when Aristotle had first raised this point in the previous chapter (185ᵃ20–32) his thought had been that 'is' applies to items in every category. This appears to be his thought here too. For he goes on to say that in any case the conclusion would not follow even if 'what is' did have one meaning, and he illustrates this by changing the example from 'what is' to 'what is pale' (τὸ λευκόν), which he assumes *does* have one meaning. If so, this is presumably because 'pale' does fall within the categories, whereas 'is' does not.

His point that it does not follow that there is only one pale thing is made in two ways. First, the various pale objects need not be one by continuity (or in any other way, we may add); secondly, 'the pale' need not be one in definition (τὸ λευκὸν ἓν τῷ λογῷ). This is because being for a pale thing (τὸ εἶναι λευκῷ) is not the same as 'being for what has received it' (τὸ εἶναι τῷ δεδεγμένῳ). Now it is clear that the main contrast he wishes to draw here is between the attribute pallor on the one hand, and the things that possess that attribute on the other, and he is noting that the one phrase τὸ λευκόν can be used of each. (Cf. *Metaphysics Z*. 6, 1031ᵇ22–8.) Since this is an ambiguity in the Greek phrase that no English expression will reproduce, we might feel that Aristotle is failing to use the phrase with just one meaning, as he had promised to do. Indeed, when the same distinction is drawn in *Metaphysics Z*. 6 he there *does* call it a case of two meanings (τὸ διττὸν σημαίνειν, ᵇ23). But evidently that thought does not strike him here. So we may put his position in this way: even though 'pale' has just one meaning, nevertheless 'the pale' need not be one, (*a*) because there may still be many pale things, and 'the pale' is used to generalize over all of them, and (*b*) because 'the pale' is also used to name the attribute pallor.

In each case he adds an *explanation* for these claims which we should not accept without reservation. He says that the various pale things are many because they are not continuous with one another, whereas in fact they might all *be* continuous with one another (like the left and right halves of a single sheet of paper) and yet still be many. He also says that pale things differ from pallor 'in definition' (ᵃ28), or 'in being' (ᵃ31),[9] or as I shall say 'in essence', and this too is problematic. His overall doctrine, of course, is that being pale never is the essence of the things that are pale, for they always have something else that is their essence (e.g. being a man, being a leaf, being a stone).[10] But an important part of the reason why being pale is not (included in) the essence of a man who is pale is just that he and his pallor may be 'separated', i.e. he may continue to exist without continuing to be pale. If, as Aristotle claims at ᵃ29–31, his argument here is not relying on anything about separability, then one wonders quite what his ground is for saying that pallor and the things that are pale do not have the same definition.[11] But these cavils are of no real importance. For the main claims

[9] No doubt this is also what he meant to say at ᵃ29, though we should have to read not λευκῷ but τῷ λευκῷ in order to get that point correctly stated.

[10] As it is put at *Post. An.* I. 22, 83ᵃ31–2: 'nothing can be pale without also being something else [as it were, *first*].'

[11] No doubt pale things have matter whereas pallor itself does not, so they cannot be identified. But it does not follow that they do not have the same definition, for Aristotle's *usual* view is that the matter of a thing does not form part of its definition. (But *Metaph. Z*. 11 clearly qualifies this view.)

of this section are clearly correct, even if one might raise doubts about the reasons given to support them.

The main claims are that, even if 'pale' does always mean the same, still (*a*) there may be many pale things, and (*b*) pallor itself is something else again, different from all of them. So the moral is that these same claims will hold not only for pallor, but also for being, which I henceforth rephrase as existence. (This is mainly to avoid what would otherwise be some very cumbersome phrasing in English; I am confident that it does no violence to Aristotle's thought.) Thus, for the sake of argument, let us suppose that 'exist' has but one meaning (i.e. that in which it applies to substances, presumably). Nevertheless, 'the existent' need not be one, (*a*) because many things (substances) may exist, and (*b*) because none of them will be existence. This brings us, then, to the next section of Aristotle's discussion.

186ᵃ32–ᵇ4. His suggestion is that Parmenides needs to assume, not merely that 'existent' means one thing, whatever it is predicated of, but also that it means 'what just is existent' and 'what just is one' (ὅπερ ὂν καὶ ὅπερ ἕν, ᵃ34). This is a highly obscure phrase. For the moment I leave it without elucidation. Aristotle goes on to explain why Parmenides needs this assumption. An accidental attribute (συμβεβηκός), he says, is predicated of an underlying thing, so that what existence is an accident of will not exist, since it is other than existence. This leads to the evidently impossible result that there will exist something which does not exist. To avoid this we must conclude that 'what just is existent' (τὸ ὅπερ ὂν) will not belong to anything else (sc. accidentally). For it—i.e. anything else, apparently[12]—cannot be something existent (ὄν τι) unless existence means many things, which by hypothesis it does not (ᵃ34–ᵇ4).

What understanding of the phrase 'what just is existent' will legitimize this reasoning? Well, it is clear that Aristotle is supposing that what is other than existence does not exist. Granted, then, that an accident is always other than the subject it is predicated of, it then follows that existence cannot be an accident of anything. For if it were then that thing would both exist (since existence is truly predicated of it) and not exist (since it is other than existence), and this is impossible. This is just the result that Parmenides requires: we have ensured that there is only one thing by taking that thing to be existence itself. And the idea flows naturally enough from the previous paragraph, which had pointed out that the phrase 'the existent' had two uses, (*a*) to generalize over existent things, and (*b*) to name existence itself. What we are now supposing is that it has only the second of these uses. How the Greek words τὸ ὂν σημαίνει ὅπερ ὂν should be construed if they are to convey this thought is not obvious, but I suggest that the idea is this: existing means being identical with existing.[13] In that case the noun

[12] One expects αὐτό in ᵇ2 to refer back to τὸ ὅπερ ὂν just before it, but then the reasoning is lost. So I take it that the antecedent must be ἄλλῳ in the preceding line.

[13] The thought is that ὄν and ὅπερ ὄν are to be equated, so one might expect Aristotle to say: τὸ ὂν σημαίνει τὸ ὅπερ ὄν. This is not *quite* what our text does say, but it is near enough. I remark that

phrase τὸ ὅπερ ὄν translates as 'what is identical with existence', which of course is a mere periphrasis for the plain phrase τὸ ὄν taken as meaning 'existence'. But the idea is that to say of anything that it exists is the same as to say of it that it is identical with existence, from which it will indeed follow that only existence exists.

So far, then, Aristotle has been setting out a position which seems to deliver the result that Parmenides wants. I presume that he cannot seriously mean to suggest that Parmenides himself *did* think in this way. Taken as an attempt at a historical reconstruction, the suggestion is obviously anachronistic, for of course Parmenides was not at all concerned with such abstract objects as existence itself, as opposed to what has it. (One might well suggest that Plato's complaint, in his discussion of Parmenides at *Sophist* 244b–245c, was that he *ought* to have been so concerned.) But once these abstract objects are brought into the discussion, then Aristotle's suggestion does perhaps have some appeal: if you really do want to hold that only one thing exists, then the best way is to suppose that that one thing is existence itself. But in the next section Aristotle tries to argue that even this suggestion will not do.

186^{b}4–12. He tries to use the result of the previous paragraph to argue for the evidently impossible conclusion that the one thing that exists, namely existence, will also not exist (^{b}10). He can then conclude, contra Parmenides, that existence must after all mean more than one thing (^{b}12). But I see no way of understanding the argument that he now offers as a correct argument.

He begins with the assumption that something, say pallor, is an accident of what just is existence. It is not clear why he feels entitled to this assumption. On the face of it, if the hypothesis is that only existence exists, then one would expect it to follow that there is nothing else which could be an accident of it. But the argument does depend upon this assumption. For we begin with the supposition that it is true to say that what just is existence is pale, and then it is claimed that pallor means a thing that does not exist, from which we are asked to infer that what just is existence does not exist (^{b}10–11). The simplest diagnosis seems to be this. Aristotle is entitled to the claim that pallor is non-existent, for this follows from his assumption that pallor is an accident of existence, and hence other than existence, whereas only existence exists. (The argument that he himself offers in ^{b}6–9 is a little more roundabout than this, but it has the same effect.) So in one sense it is true that 'pallor means a thing that does not exist', i.e. in the sense in which this just says that pallor is non-existent. But Aristotle then takes it in a different sense, as if it claimed that pallor is non-existence, so that to say that something is pale is to say that it does not exist. But quite clearly the second sense

elsewhere where Aristotle employs his peculiar ὅπερ locution (e.g. *Post. An.* I. 22, *passim*) it is again attractive to suppose that he means it to import the notion of identity. I have explored this in my (2004).

does not follow from the first. I do not see how the reasoning that Aristotle gives can be rescued from this charge.[14]

Charlton (1970: 61) offers an ingenious way of strengthening the argument so that its conclusion really does follow. We began with the idea that, in order to maintain his position, Parmenides must claim that existence means only one thing, which is then taken to imply that only existence exists. Charlton now suggests that, by parity of reasoning, Aristotle is entitled to assume that pallor means only one thing, and hence that only pallor is pale. If this is granted, then we can fairly argue that if existence is pale then existence must be pallor, but it is agreed that pallor does not exist, so existence will not exist. However, one has to admit that the extra premiss which is here being relied upon, i.e. (in Aristotle's language) the premiss that τὸ λευκὸν σημαίνει ὅπερ λευκόν, is not to be found anywhere in the text that we have.

In any case, Aristotle's refutation depends upon the initial assumption that something (say pallor) will be an accidental attribute of existence, and I see no reason why Parmenides should be expected to grant that. I see no *self-contradiction* in the hypothesis that only existence exists.

4. THE CONCLUDING SECTION

The overall structure of the final section of this chapter, i.e. 186^b12–187^a11, is desperately obscure. But there is one stretch of detailed argumentation within it (i.e. 186^b14–35), which it will be convenient to take first. This apparently returns to a theme introduced at the start of the discussion, namely how it is possible for the same thing to be both one and many (185^b25–186^a3). The example here is: how can one thing, a man, also be the two things, an animal and a biped?[15]

In this discussion it seems to me quite clear that the phrase τὸ ὅπερ ὄν is used in a quite different way from its use in the previous section. Sticking to the idea

[14] The final step of the argument ($^b11$–12) is not altogether clear. I reconstruct it thus. The hypothesis that pallor means a thing that does not exist has led to a contradiction, and so must be rejected. Hence pallor means (i.e. is) a thing that does exist. But, *retaining* the assumption that pallor is an accident of existence, pallor and existence cannot be the same thing. So there will be two things that exist, namely pallor and existence, and hence existence cannot mean just one thing. Admittedly, on this reconstruction one expects not quite what our text says, namely ὥστε καὶ τὸ λευκὸν σημαίνει ὅπερ ὄν, but rather: . . . σημαίνει ὄν τι. I am now inclined to think that this discrepancy is not important. But one may be sympathetic to Natorp's conjecture: . . . σημαίνει [καὶ] ὅπερ ὄν. The meaning will then be 'It (i.e. τὸ ὅπερ ὄν, subject imported from the previous sentence) will mean (i.e. be true of) both pallor and itself.'

[15] Throughout the Logical Works Aristotle takes man to be defined as an animal that walks and has two feet, and the example given here is his standard example elsewhere to introduce the problem of how a *definition* can be a unity. (This is mentioned as a problem at *De Int.* 17^a13–15 and *Post. An.* 92^a29–33; there are attempts at solutions in *Metaph. Z.* 12 and *H.* 6.) Here Aristotle does not introduce his problem as one that specifically concerns definitions, though it is clear from his discussion that that is how he is thinking of it.

that the word ὅπερ introduces the notion of identity, I would say that in the present section what is ὅπερ ὄν is construed as what is identical to an existent thing, whereas in the previous section it was construed as what is identical to existence. This new use is directly parallel to what one finds elsewhere in Aristotle, where (as here) no distinction is drawn between ὅπερ ὄν and ὅπερ ὄν τι.[16] Just as τὸ ὄν can mean either 'an existent thing' or 'existence', so also with τὸ ὅπερ ὄν. In the present passage the phrase is applied to just those things that Aristotle regards as existing 'in their own right' and 'independently of anything else', namely the substances. On his account, (a) man is a paradigm example of this, and (a) man is by definition both (an) animal and (a) biped. He argues here that since the thing we begin with is a substance, so also are the two things into which it divides. For the alternative is that they are accidental attributes, and that alternative is what our passage aims to refute. I briefly run through the argument.

If being an animal, and being two-footed, are accidents, then they are accidents either of man or of some other underlying thing (^{b}17–18). But an accident is either an attribute that can both belong and not belong to its subject, or an attribute which has in its definition the subject to which it belongs.[17] (The first is illustrated by the attribute of being seated, the second by the attribute of being snub.) Moreover, if one thing occurs in the definition of another, then that other does not also occur in the definition of the first. Since, then, two-footedness does occur in the definition of man, man cannot also occur in its definition, so it is not an accident of man in the second sense. But also, there cannot be men who are not two-footed, so it is not an accident of man in the first sense either. (And here Aristotle tacitly assumes that the same will apply to being an animal.) If, then, these are accidents—i.e. if each is not what just is a certain existent—they must be accidents of something else. (And here Aristotle assumes, without evident warrant, that both will be accidents of the same subject, and so he concludes:) Man, then, will also be an accident (sc. of that same subject). But that he rejects as impossible, for we may assume that what just is existent is not an accident of anything. This concludes the discussion.

We may say that the point that this paragraph argues for is that the 'logical constituents' of a substance—e.g. as 'animal' and 'two-footed' are logical constituents of 'man'—must themselves be substances. And the main and controversial principle being relied upon is that, since these constituents belong necessarily to the substance, they cannot be counted as accidents of it, and hence are not accidents at all, and so must be substances themselves. In view of the many uses that Aristotle gives to the notion of an accident, it is not particularly surprising to find the word being used here for anything not in the category of

[16] For a fuller discussion see my (2004).

[17] In Aristotle's more usual way of speaking, an attribute such as snubness, whose definition must include a mention of what it applies to (namely noses), is called an attribute 'in its own right' (καθ' αὑτὸ), and *contrasted* with an accident. See e.g. *Post. An.* I. 4, 73^{a}34–^{b}5; *Metaph. Δ*. 18, 1022^{a}29–32; *Z*. 5 *passim*.

substance. But the argument as a whole carries no conviction, for its leading principle is entirely without foundation. We are given no reason to suppose that an attribute which is not in the category of substance cannot be a necessary attribute of what is in that category, and this claim is not at all plausible. But Aristotle does here put it forward, apparently *in propria persona*, so we can only suppose that he here means to endorse it.[18]

But let us now come back to the question that I noted at the outset. How is this little passage of argument supposed to fit into its context?

The criticism of Parmenides that we considered earlier ends with a separate two-line argument: if what exists is what just is existing, then it will have no magnitude; for if it did it would have parts, and they would each have a different existence (186^{b}12–14). Here the expression ὅπερ ὄν is presumably to be understood as 'what just is existence', and the point being made is that if only existence exists then it can have no parts and therefore no size. (This is a criticism of both Parmenides and Melissus, since the one claimed that what exists is limited, and the other that it is unlimited, and Aristotle thinks of each of these as implying that it has size.) If so, then—as Ross supposes—this little argument belongs with what has preceded it, rather than what follows, since as we have seen what follows uses ὅπερ ὄν in a different sense. But it does reintroduce the earlier question of how it can be that what is one (whole) is also many (parts), and that is the theme treated next.

For what comes next is the argument that we have just looked at in some detail, which is introduced as showing that what just is an existent thing (i.e. a substance) is divisible into other things which also just are existing things, a point which it claims 'is also clear from argument'.[19] This argument pursues its course, as we have described, but then it ends quite unexpectedly with 'So is the universe made of indivisibles?' (^{b}35). Then in what seems to be a concluding section at ^{a}1–10 Aristotle begins by saying that some people gave in to both arguments. To the argument that all things will be one if 'the existent' means only one thing, they responded by claiming that there was such a thing as 'the non-existent' (i.e. empty space); and to the argument from dichotomy they responded by positing atomic magnitudes.[20] Then he urges once more that the first response was

[18] In my (2004) I have discussed whether *in the Logical Works* Aristotle does or should count the differentia of a substance as itself a substance. (It appears to me that, in the end, he does, though he should not have done.) If we look elsewhere, especially in the *Metaphysics,* we find no coherent doctrine. (For example, *Z*. 10 claims that the constituents of a definition are prior to the thing defined, and since we know that no non-substance can be prior to substance it must follow that the constituents of the definition of a substance are themselves substances. But *Z*. 13 claims that no substance can have constituents that are substances.)

[19] καὶ τῷ λόγῳ φανερόν (^{b}15). In what *other* way is it also clear? Or could this mean that it is clear *even* from [consideration of] a definition?

[20] ἄτομα μεγέθη. It is not clear whether this literally means atomic, i.e. indivisible, *sizes*, or whether it is a way of talking of atomic *bodies*, namely bodies of that size. In any case, Greek thought constantly associated the two.

unnecessary, and although he does not say so here one can be sure that he felt the same about the second. But the extraordinary thing is that he writes as though he has just been *giving* us 'the argument from dichotomy', and perhaps giving us a criticism of it too, yet evidently he has done no such thing.

This argument from dichotomy must surely have been an argument of Zeno's (or a descendant of one of his), pointing to difficulties in the idea of infinite divisibility. In book VI of the *Physics* Aristotle uses this title for the first of Zeno's well-known paradoxes on motion (239^b18–20, referring back to $^b11$–14), and although his discussion of this argument in book VI does not explicitly mention its connections with atomism, still book VI is as a whole a sustained critique of all kinds of atomism.[21] Alternatively the title would apply well enough to (one version of[22]) an argument about the infinite division of an object that Aristotle wrestles with in book I of the *De Generatione et Corruptione*, at 316^a14–317^a12, and which he does explicitly recognize as an argument for atomism (316^a11–14, $^b14$–16, $^b32$–317^a1, $^a12$–15). One of these, or perhaps some other variation on this general theme, must be what Aristotle is referring to as a motive for atomism. But clearly our text contains no statement of any such argument. Ross has noted an interesting hint that Aristotle was *thinking* of such an argument at the beginning of our passage, i.e. 186^b12–14. For he says there that if only existence exists then what exists will have no size, since each of its parts would have to have a different being. But the Greek for 'each' here is ἑκάτερος, which of course is used for each of *two*, not each of an indefinite number. Why should Aristotle be thinking of the existent as having just *two* parts, if it has size? Well, perhaps because he is already beginning to think of the dichotomy, which first divides a thing into two parts, and then divides one or both of these into two parts again, and so on indefinitely. But although a dichotomy may perhaps be hinted at when our section begins, at 12–14, and is apparently recalled when it ends, at $^a1$–3, what our text gives us in between is not at all the kind of argument to which Leucippus or Democritus might have responded by positing atoms.

One might naturally expect Aristotle to be thinking of Zeno in this chapter. Admittedly it is I who have written as my title 'Aristotle on the Eleatics', and Aristotle himself never uses the word 'Eleatic' at all, so far as I know, and anyway not in our sense in which it is short for 'Parmenides, Zeno, Melissus'. Nevertheless, the tradition which sees a close association between Parmenides and Zeno is certainly as old as Plato's *Parmenides*, and probably much older still. (After all, those two did come from the same place, Elea, as Melissus did not.) So it would be perfectly natural for Aristotle to plan a chapter in which he first discusses the arguments of Melissus (as at 186^a10–18), then what he took to be the basis of Parmenides' position (as at 186^a22–$^b12$), and finally Zeno's contribution. Then the chapter could end with a mention of what was, in his eyes,

[21] I have defended this description of book VI in my (1991), Essay 10 in this volume.

[22] I have distinguished the versions in my (1988). I also note there that Aristotle does not have the conceptual resources to deal adequately with this argument. (See pp. 186–8 in this volume.)

the harmful legacy of these three Eleatics together, namely the rise of atomism. But the text that we have fails to contain the expected section on Zeno, though as we have seen it does contain *hints* of such a section. *Perhaps* this is because our text has been mutilated, and the section on Zeno has been lost. But I think it is also a possibility that Aristotle's own text was missing this section. He had *intended* to put in something on Zeno, and when he wrote his conclusion he wrote as if he *had* done (for it was there in his thought), but in fact he had been distracted by the rather different argument that we now find at ^{b}14–35. This concerns, not a division into spatial parts, but what one may call a division into *logical* parts, i.e. into 'parts' mentioned in a definition.

Does he think of it, then, as the logical analogue of a Zenonian argument from dichotomy about spatial parts? If so, then he must be supposing that every substance has logical parts which are substances, so that an infinite regress is involved. But it seems very improbable that he believed in the correctness of this infinite regress, for that would have the consequence that a definition (when fully spelled out) must be infinitely long, and that is a suggestion that he consistently rejects. (See e.g. *Post. An.* 82^{b}37–83^{a}1; *Metaph. H.* 3, 1043^{b}35–6.) It follows that there must be some 'logical indivisibles', and so in this somewhat peculiar sense of the word the question 'So is the universe composed of indivisibles?' must be answered 'Yes'. On the other hand, when we take the question in its more normal sense, as asking about indivisible bodies such as the Atomists believed in, then it is quite clear that Aristotle's own answer to it is 'No'. What, then, is the relation between the Zenonian dichotomy that we expect to find in this place, and the argument about 'logical division' that is actually there? Was Aristotle perhaps toying with the idea that an Atomist might wish to argue that, *just as* a logical division must end with indivisibles, so the same must be true of a spatial division? (If so, it is an argument that he himself would be bound to object to.) Well, that is a mere speculation, attributing to Aristotle's thought a great deal more than our text actually says, but I know of no better solution to this conundrum.

7

Aristotle, Zeno, and the Potential Infinite

1. ARISTOTLE ON THE POTENTIAL INFINITE

Aristotle is generally thought to have held that nothing is *actually* infinite though several things are *potentially* infinite. It is certainly true that Aristotle often expresses his view in just this way, but it seems also to be true that he recognized that this was to some extent a misleading way of expressing it. Consider, for instance, the words with which he first introduces the view, in *Physics* III.

In chapter 5 he has been arguing that there cannot be any physical body which is infinitely large, though he expresses this simply as the claim that no physical body is, or could be, infinite. In chapter 6, however, he observes that we can hardly deny the infinite any existence at all, for that would imply that time had a beginning and an end, that magnitudes were not always divisible into further magnitudes, and that the number series came to a stop. From this he infers that we need to distinguish the senses in which the infinite may be said to exist or not to exist, and continues:

> A thing may be said to be either potentially or actually, and a thing may be infinite either by addition or by division. Now we have said that no magnitude is actually infinite, but magnitudes are infinite by division (for the thesis that there are indivisible lines is easily refuted), and therefore they must be potentially infinite. But their being potentially infinite must not be understood as implying that they will (may?) at some time be actually infinite in the same way as what is potentially a statue will (may?) at some time be a statue. For *being* has many senses, and the sense in which a thing *is* infinite is the sense in which there *is* a day or a contest, namely by one thing coming into existence after another. For indeed in these cases too we may distinguish between potentiality and actuality: the Olympic games (may) exist both in the sense that they are able to take place and in the sense that they are taking place. (206^a14–25)

What is clear is that Aristotle begins by claiming that a magnitude such as a length is potentially but not actually 'infinite by division', and then goes on to expound the sense of this claim. It may be noted first that although he writes as if his claim were a conclusion from argument it is in fact nothing of the sort. For what has been said previously is that no magnitude is *infinitely large* (either

Reprinted from *Proceedings of the Aristotelian Society*, 73 (1972–3), 37–51.

actually or potentially), but being infinitely large is evidently a case of being 'infinite by addition' (cf. 206^{b}20–7) and so has just been distinguished from being 'infinite by division'. In fact no part of the preceding discussion has any tendency to show that a line cannot be actually infinite by division, and no part of the following discussion seems to offer any *argument* for this claim either. In the end I think it has to be admitted that this is simply a lacuna in Aristotle's account, and a lacuna of some importance, but for the moment let us leave the grounds for Aristotle's claim on one side and consider its elucidation.

Normally, to say that something is potentially ϕ is to imply that it is possible that it should be actually ϕ, and it might seem that Aristotle is intending to deny this implication in the case of infinity. If this were so, he would clearly be using the word 'potentially' in an unusual sense, but in fact it is not the sense of 'potentially' that he mentions as unusual but the relevant sense of 'is': the relevant sense of 'is' is the sense in which a process such as a game may be said to 'be', namely by one thing coming into existence after another. Now in the ensuing discussion it does very often seem that when Aristotle talks of a thing being 'infinite by division' it is indeed a *process* that he has in mind, namely the process of continually dividing the thing (cf. e.g. 206^{a}27–9). But a process, he here states, may be said to exist actually at any time at which it is going on, and he does *not* deny that the process of continually dividing a finite line can in this way exist actually (206^{b}12–14). But it seems equally clear that the sense in which a finite line is potentially infinite by division is again that the process of continually dividing it exists potentially (i.e. *could* exist), and so it turns out in the end that in the sense in which a line is potentially infinite by division it may *also* be actually infinite by division. In that case, what has become of the claim that the infinite is potential but *not* actual?

I think it is clear that what Aristotle mainly had in mind when he first made this claim is just that the process of dividing a line into infinitely many parts is one that cannot be *completed*, and one can indeed understand his contrast with the potential statue in this light. It may well be that what is potentially infinite by division *actually is being* infinitely divided, but it cannot be that it *actually has been* infinitely divided; by contrast a lump of bronze that is potentially a statue may certainly *be becoming* a statue but it may also *have become* a statue, and it is this latter that we mean when we say that it *is* a statue. (And one can see how Aristotle could say that this contrast, which I have just drawn by using different tenses, is a contrast in the relevant senses of 'is'.) So it appears that the claim that Aristotle is making is just the claim that an infinite process is a process that cannot be finished—for, one might ask, how could one come to the end of that which has no end?

Now for the moment we may grant Aristotle's views on infinite *processes*, but it is relevant to recall that infinity is not usually considered to apply only to processes. For instance, it would usually be held that there are infinitely many numbers (as Aristotle has already remarked), and that there are infinitely many points on a finite line, and neither of these cases of infinity appears to be a process in the relevant sense, that is, a series of occurrences which take place one after the other. So when Aristotle claims that these infinities are potential and not actual he cannot easily be understood as claiming that they are processes which cannot be completed. Yet I think that this *is* very close to what he is claiming, for—to consider only the points on a finite line—he does hold that these are dependent on occurrences. In fact he distinguishes between the actual and the merely potential existence of a point in such a way that a point is not said to exist actually until it *has been actualized*, so that there could be an actual infinity of points on a line at one time only if infinitely many points of the line had been actualized by that time. But, assuming that one cannot actualize infinitely many points *all at once*, this must involve the completion of an infinite process of actualizations, and that we have (for the moment) agreed to be impossible. Thus it is that the infinity of points on a line turns out not to be an actual infinity (though it still is a potential infinity, on the ground that there could exist a process of actualizing points on the line which was an infinite process).

The first major difficulty with this view is evidently to understand the notion of actualizing a point, and the second is to see why Aristotle was led to the view in the first place. For while it does at first sight seem reasonable to claim that an infinite or unending process cannot ever be completed, there is no such initial plausibility in the claim that an infinite totality could exist only as the result of a completed infinite process. Why, then, does Aristotle wish to deny the actual existence of infinitely many points on a line, and how does he suppose that a point is brought into actual existence? It seems to me that the answer to both these questions is to be found in his treatment of Zeno's problem, to which I now turn.

2. ARISTOTLE ON ZENO

The heart of Zeno's problem may be stated in this way. Suppose we have a body (which we may call Achilles) which moves a finite distance from a starting point to a finishing point (which we may call the tortoise). Now the distance over which Achilles has to travel to reach the tortoise will contain infinitely many points, and any infinite series of such points will mark out the distance into infinitely many discrete parts. Consequently Achilles has to traverse infinitely many parts of the distance, one after the other, before he has traversed the whole distance. But this is as much as to say that before he can reach the tortoise he must have completed a series of infinitely many different tasks, and we have just agreed that this is impossible. So Achilles cannot catch his tortoise after all.

Aristotle's answer to the problem, when stated in this form, is given in *Physics* VIII. 8, 263^a4–$^b9$, and this answer relies heavily on the view that a line does not actually contain infinitely many points. Aristotle still maintains (i) that it is impossible to have completed a series of infinitely many actual tasks, and he agrees that traversing an actual part of a line would be an actual task. But he also holds (ii) that an actual part of a line must be bounded by actual points, so Achilles would have to traverse infinitely many actual parts of the line only if he had to pass through infinitely many actual points. However, (iii) the points that Achilles passes through cannot be expected to be actual points unless Achilles in some way actualizes them as he passes through them, and *merely* passing through a point is not sufficient to actualize it. In fact (iv) a point in a line may be actualized by a body's coming to rest at that point, or by a division actually being made at that point, or indeed by the point's being merely counted; and an infinite series of tasks such as these *cannot* be performed. So finally (v) what we do think Achilles can do, namely to move continuously until he reaches the tortoise, is not after all an infinite series of tasks, but just one task, for it is not actually split up into component tasks. One could specify a task that splits up into an infinite series of component tasks, by requiring Achilles to actualize some infinite series of the intervening points in some way, but that would be specifying a new task and one that Achilles could *not* perform. (For instance one could require Achilles to move $\frac{1}{2}$ the distance in $\frac{1}{2}$ a minute and then rest for $\frac{1}{2}$ a minute; then to move $\frac{1}{4}$ of the distance in $\frac{1}{4}$ of a minute and then rest for $\frac{1}{4}$ of a minute; and so on. Evidently if Achilles could complete this infinite series of tasks he would reach the tortoise in 2 minutes, but Aristotle says—and common sense seems likely to agree with him—that Achilles *cannot* complete *this* series of tasks.)

Now it seems to me that this solution to Zeno's paradox is worth some consideration, for at least at first sight it does give the right results concerning what Achilles can and cannot be expected to do; he can be expected to pass through infinitely many points but he cannot be expected to 'actualize' them, where by 'actualizing' a point we mean roughly *doing* something at or to the point which singles it out in some way from its neighbours. That this is the right sort of way to understand the notion of actualizing a point may be seen by considering similar Zenonian problems in other areas. For instance, having shown by his well-known arguments that nothing ever moves, Zeno might have completed his dilemma by arguing that nothing ever rests either. For, for Achilles to remain at rest for a full minute he must first remain at rest for $\frac{1}{2}$ a minute, then for the next $\frac{1}{4}$ minute, then for the next $\frac{1}{8}$ minute, and so on. Thus he must have completed an infinite series of tasks of resting before he can be said to have remained at rest for his minute, and that we have said is impossible. But presumably the appropriate Aristotelian reply to this problem is that one does not actualize an instant of time by simply enduring through it; to actualize an instant one has to *do* something *at* it which in some way singles it out from its neighbours, and that condition is not fulfilled by simply remaining at rest during a period in which it falls.

This characterization of the notion of actualizing a point is still very vague, of course, but I am inclined to think that it is at least definite enough for us to see that it does not after all contain the solution to Zeno's problem. First, it is relevant to notice a very obvious objection, namely that when Achilles passes through any intermediate point in his journey to the tortoise then there is of course something which he does at that point, viz. he reaches or arrives at it. Now in fact Aristotle goes so far as to argue that this is not after all true. For, he says, if Achilles arrives at the point in his journey then clearly he also leaves the point, for otherwise he would remain there and so never reach the tortoise. But at the time when Achilles has arrived at the point then he is at the point, whereas at the time when he has left the point he is not at it (but beyond it), so the time when he has arrived and the time when he has left are not the same time, and consequently they must be separated by an interval. But during that interval Achilles must evidently be at the point, since he has arrived at it but has not yet left it, and so we may conclude that if Achilles reaches or arrives at any point in his journey then he also rests at it. Hence finally when Achilles is travelling continuously, as we are supposing he is, we must not say that he reaches or arrives at any point which he passes through. (262^{a}19–263^{a}3.)

Well, this argument is of course fallacious, for if *t* is the time when Achilles arrives at the point, then all times later than *t* are times when he has left it, but there is evidently no interval of time between *t* and all times later than *t*. But it is interesting that Aristotle produces the argument, for this does perhaps indicate that he sensed it as an objection to his notion of actualizing a point that merely arriving at a point ought to be enough to actualize it. However, I do not want to press this objection, which may perhaps seem rather captious, but to turn to the idea of reversing one's direction at a point, which was in fact Aristotle's ultimate target in this argument. For one of the reasons why he argued that arriving at a point, and leaving the point, involve remaining at the point for at least some time, was precisely that he wanted to show that when I reverse my direction at a point then I *must* rest at that point for some interval of time, since in this case we cannot avoid speaking of my arriving at the point and leaving it. (And he even thinks that observation assures us of this! 262^{a}17–19.) Now this is clearly a mistake, because, for instance, the point which is the limit of a projectile's upward motion and the start of its downward motion is a point at which the projectile reverses its direction without remaining at the point for any period. But it seems absolutely clear that a point at which one reverses one's direction of motion is a point at which one *does* something that singles that point out from its neighbours, and hence a point which is actualized. And it seems to me that this creates a difficulty for Aristotle's theory of the potential infinite.

3. ZENO'S PROBLEM RESTATED

I shall develop this difficulty with the help of an example, and I take as my example a theory to explain how it is that an ordinary rubber ball bounces on a rigid and unyielding surface. In outline the theory is that when the falling ball strikes the surface, its momentum carries it down yet further, thus compressing the rubber on the side of the ball in contact with the surface and deforming the ball from its natural shape. The ball, however, is elastic, and so resists deformation, with the result that the downward motion of the ball is eventually halted and the deforming process comes to an end. Thereupon the reverse process begins: the ball reverts to its natural shape, exerting pressure on the surface and thereby raising itself once more, and the upward momentum which it thus acquires carries it away from the surface and so brings about a further bounce.

More exactly, let us call the shape which the ball would assume when remaining at rest on the surface its rest shape, and the position it would then assume its rest position. (The rest shape will be slighly deformed, since the ball's own weight will compress the rubber next to the surface to some extent). Then the theory is that the ball not only accelerates during its downward fall but continues to accelerate downwards even after it has made contact with the surface until it has been deformed into its rest shape and reached its rest position. After it has made contact with the surface the rate of acceleration downwards will of course be diminished, but it will still be positive, for the gravitational attraction responsible for this acceleration will be greater than the opposing upwards force due to the ball's resistance to deformation, until the rest position is reached. And this must be so, for what holds the ball in its rest position when it is at rest is that at that point the downward force of gravity and the upward force due to the ball's resistance to deformation just balance, and therefore at any higher point the force of gravity will be greater than the ball's resistance, and at any lower point the ball's resistance to deformation will be greater than the force of gravity. Consequently the ball's downward momentum will be at its maximum at the time when the ball passes through its rest position on the way down, and by parallel considerations the ball's upward momentum on the next bounce will be at its maximum again at the time when the ball passes through its rest position on the way up. Finally, we assume that the ball does not acquire or lose momentum by fits and starts; all such changes take place strictly continuously.

I have elaborated this theory in some detail because I want to make it quite clear that the theory is an entirely plausible one, and does not involve anything unintelligible. But its point of course is that according to this theory once a ball has started bouncing there is no bounce that is its last bounce. Actually we probably would not say that there had been a *bounce* in the ordinary sense unless the ball has actually left the surface, and it is not a consequence of the theory that every time the ball meets the surface it must leave it again. But it is a consequence of the theory that every time the ball sinks below its rest position it must rise above its rest position again, simply because if it sinks below its rest position it is deformed beyond its rest shape, and its resistance to deformation will then raise it once more to its rest position, and in such a way that it acquires its maximum momentum upwards at the time when it reaches its rest position. Inevitably, then, it will continue to rise further above that position, being carried on by this momentum, until its upward motion is eventually brought to a halt by the deceleration due to gravity. So, as I say, every time the ball sinks below its rest position it will rise above it again, and by parallel considerations every time that it rises above its rest position it will sink below it again, and consequently no oscillation about its rest position will be the last oscillation.

But it would be an error to conclude from this that once our ball has started bouncing (or oscillating) it must continue to do so for ever. *That* is not a consequence of our theory at all. It is an obvious fact of experience that when an ordinary rubber ball is set to bounce its bounces become successively smaller and smaller, taking less and less time each, and this fact is entirely consistent with our theory. Indeed, a moment's reflection is sufficient to show that it is quite consistent with the theory to suppose that with some balls each bounce takes only half as long as the preceding bounce, so that the times taken in successive bounces are in the continued proportion

$$1, \tfrac{1}{2}, \tfrac{1}{4}, \tfrac{1}{8}, \tfrac{1}{16}, \ldots$$

and all the bounces are completed in a finite time. And of course it is equally consistent to suppose that the times taken in successive bounces are in some different proportion, for instance

$$1, \tfrac{1}{2}, \tfrac{1}{3}, \tfrac{1}{4}, \tfrac{1}{5}, \ldots$$

so that the bouncing goes on for ever. The theory I have been giving is neutral between these two alternatives, and which kind of series proves to be realized in any given case must wait upon empirical investigation. What I am claiming is simply that the first alternative is a perfectly good possibility, so that it is quite possible that an infinite series of bounces—which is one sort of infinite process—should have been completed.

If all this is admitted, then it seems quite clear that Aristotle's solution to Zeno's problem must be rejected. For Aristotle, wishing to uphold his claim that an infinite series of tasks cannot be completed, was led to argue that Achilles does not traverse infinitely many parts of the line joining his starting point and his goal, on the ground that there are not infinitely many points on that line marking it out into parts. But he was able to deny the (actual) existence of these points only by denying that they had been actualized, for Achilles was not supposed to have done anything at or to the points as he passed them. However, this reply is surely not available to the revised version of Zeno's problem that I have just put forward, for the ball's motion is certainly divided into infinitely many parts by the infinitely many points which mark the top of each bounce, and these must surely be admitted to have been actualized in the course of the bouncing. I conclude, then, that Aristotle does not after all have the right solution to Zeno's problem.

4. FURTHER CONSIDERATIONS

But perhaps the argument so far is over-hasty, for there is certainly a general reluctance to admit that an infinite series of tasks can be completed, and we have not yet uncovered the source of this reluctance. Indeed it has been argued that such a completion is impossible on logical grounds, but these arguments do not stand up to scrutiny.

The argument which I supplied for Aristotle earlier—viz., that it must be impossible to come to the end of that which has no end—need not detain us long, since it is fairly evident that it rests on an equivocation on the phrase 'come to the end of'. Certainly an infinite series has no last member, and therefore it is indeed impossible to *come to the last member* of the series. But it by no means follows that it is impossible to *finish* the series, i.e. to come to a state in which no member remains outstanding. From the fact that there is no *last* member it does not follow that we cannot perform *every* member.

The other arguments usually advanced here are arguments purporting to show that we cannot give a consistent account of the state which results when every member of the series of tasks has been performed. In the case of the bouncing ball (as so far described) such arguments cannot get started, because it is perfectly clear that in the final state of the system the ball will simply be at rest on the surface. But suppose we were to harness our ball to some kind of switch mechanism which will be in one of its two states during the performance of one bounce and in the other during the next bounce. Then, it will be argued, the switch cannot consistently be supposed to be in either of its two states when the infinite series of bounces has been completed. But the reply to such arguments is generally that there is no such inconsistency as they allege, and if there is anything puzzling here it is just that *neither* of the two incompatible end-states is inconsistent with the initial specification.

In more detail, let us begin with an entirely fanciful switch mechanism as an example, to avoid the difficulty that any ordinary switch mechanism will have to be able to operate at arbitrarily high speeds. Let us suppose that our bouncing ball is white during its first bounce and changes colour from white to black and back again at each bounce—in fact let us specify that it changes colour exactly at the moment that it passes through its rest position on the upward journey, and that it changes colour at no other time. Now I agree that if the original account of the bouncing ball was intelligible then this further specified account is also intelligible. But, it will be argued this further specified account has become unintelligible. For since our ball can change colour only between white and black, it must be either white or black when the bouncing is finished. But if it is white then there were an odd number of bounces in all, and if it is black then there were an even number of bounces in all, and yet neither of these is possible since only a finite number can be odd or even, and the number of bounces was supposed not to be a finite number. However, a little thought will show that this argument is guilty of an illicit conversion: from the initial specification it can indeed be deduced that if the ball bounced for an odd number of times it would end as white, but it cannot be deduced that if the ball ends as white then it bounced for an odd number of times. In fact the position is that we have been given enough information to determine the colour of the ball during all the time that it is bouncing, but we have not been given enough to determine its colour when it has finished; or, put in another way, our information will give us the colour of the ball at any time when it has completed only a finite number of bounces, but it does not tell us what happens when the ball has completed infinitely many bounces.

To see this more clearly, let us attempt to deduce its final colour. If we consider the ball's first moment of rest—and I mean the first moment such that the ball is at its rest position at that moment and at all succeeding moments—then it is clear that the ball does not change colour at that moment. For we specified that it changed colour only when *passing through* its rest position (on the upward journey), and it is not passing through its rest position at its first moment of rest, but coming to rest at it. So, one is inclined to say, if it does not change colour at that moment we ought to be able to infer that it will be the same colour then as it was *at the previous moment.* But of course the trouble is that there is no such thing as the previous moment, for quite generally no two moments are *next* moments (since between any two moments there are others). The most that one could hope to infer, then, is that the ball will be the same colour at its first moment of rest as it was at *all* preceding moments within a sufficiently short interval. But here the trouble is that there is no immediately preceding interval that is short enough to ensure that the ball remains the same colour throughout it, and all methods of trying to deduce the ball's colour at its first moment of rest must therefore fail. Consequently we are at liberty to make what assumptions we like about it. We may take it to be black and no contradiction will result, and we may take it to be white and no contradiction will result. Where, then, is the difficulty?

If there is any difficulty it is just that we mistakenly suppose that the initial specification *must* suffice to determine the ball's colour at all subsequent times, and this is particularly liable to be felt as a difficulty if we are considering not my fanciful example of colour-changing but something that is recognizably a *causal mechanism*. Suppose, for instance, we imagine that whenever the pressure of the ball on the surface is* greater than its pressure when at rest on the surface it closes a circuit permitting current to flow which moves a genuine switch from one of its two positions to the other. Then to the question where the switch is at the end of the bouncing we must reply as before that we have not been given enough information to determine the matter, but in this case the reply is apt to seem quite mysterious. For in this case the lack appears not to be in the information

* This 'is' should be 'becomes'.

we have been given but in the nature of things: if the final position of the switch is not deducible from this information, then it is not determined by the causal laws which govern the operation of the switch, and so whatever happens there will have been a breakdown in causal determinism. For we may suppose that we are given the initial positions of the ball and the switch at a certain time, and that we possess a theory concerning the subsequent behaviour of the ball which allows us to predict its position and momentum at all subsequent times. We also have a further theory connecting the behaviour of the ball with the behaviour of the switch, which we would say makes the behaviour of the switch wholly dependent on that of the ball—at least, the behaviour of the switch is not determined by anything else than by the behaviour of the ball. But we now find that although we can predict the position of the switch at all times before the ball's first moment of rest, we cannot predict its position at that moment or after it. It seems clear that our inability to predict is not due to a lack of knowledge of the initial state of the system, and it does not seem at all plausible to say that it is due to a lack of knowledge of the causal laws governing the system, and if this is granted it then follows that the system is not deterministic in a Laplacian sense: whatever the final position of the switch turns out to be, it is uncaused.

My conclusion is that there is not in general any *logical* impossibility in the idea of completing an infinite series of tasks; this is actually achieved by Achilles when he catches his tortoise, and there is no difficulty in supposing it achieved by the bouncing ball. On the other hand there are cases where the idea that such an infinite series may be completed runs into conflict with our belief in causal determinism (and, incidentally, with our belief that every mechanism has a highest speed of operation), and it seems to be these cases that lead people to say that the completion is impossible. I shall end by making just one suggestion as to the fundamental difference between the cases where this conflict does arise and where it does not, though I have no space here to explore the suggestion in any detail.

If we draw a graph to represent the motion of our bouncing ball it will have the general shape

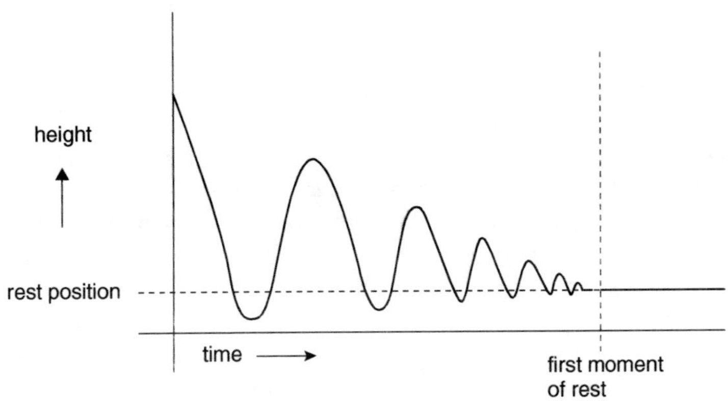

and this curve is continuous, despite the fact that there are infinitely many oscillations before the first moment of rest. However, if we now consider the motion of the troublesome switch, we may imagine this to have a pointer that is in 'up' position during odd-numbered bounces and in 'down' position during even-numbered bounces, and moves ever more rapidly between the two. In that case a graph representing its motion would have the general shape

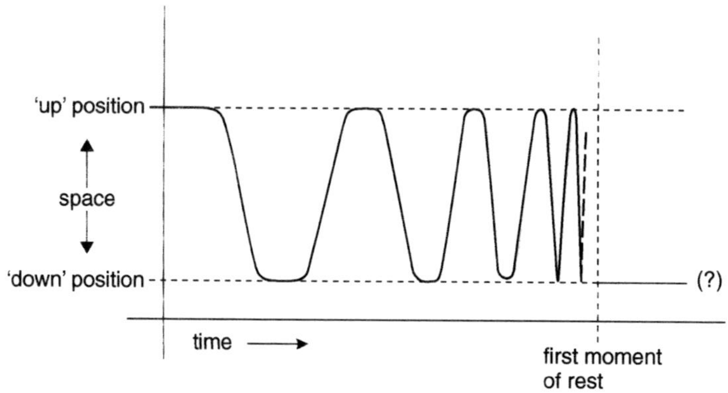

and now, wherever we suppose the pointer to be at and after the first moment of rest, the curve *must* be discontinuous at that point. There is no logical impossibility in the idea of a motion that is represented by such a discontinuous curve, but I think we do find it repugnant to our idea of causality in nature, for the state of the switch at and after the first moment of rest cannot be smoothly connected with its previous states. When a completed infinite process would involve such a discontinuity in nature, then—and perhaps only then—we tend to regard it as impossible.

Additional Note 7.1

Although this essay was written some time ago, and is in fact the earliest of the essays reprinted here, I do not wish to retract a word of it. Instead, I add two brief remarks on topics which it does not cover. As is mentioned on p. 116, Aristotle's positive discussion of the infinite begins (at the start of *Physics* III. 6) by noting *three* cases where it is difficult to deny the existence of an infinity. These are the infinity of time, the infinite divisibility of magnitudes, and the infinity of the numbers. The essay here reprinted is wholly concerned with the second of these three topics. What of the other two?

What could it mean to say that the numbers are only potentially infinite?[1] If we may extrapolate from what Aristotle says of points, this should apparently mean that numbers do not actually exist until something happens which 'actualizes' them, and there is or could be a process of 'actualizing' the numbers, one by one, which need not ever end, but which will at no time ever be completed. One asks: what kind of process could this be? How far has it got to by now? (For example, how many numbers were there yesterday, and will there be more tomorrow?) This picture seems to me to be plainly absurd, and it must embody an entirely mistaken conception of what numbers are. Nevertheless, it has its adherents even today, in those philosophers of mathematics who are known as 'intuitionists'. They do claim both that numbers are 'mental mathematical constructions' and that they form only a 'potential' infinity, since the number of mental constructions which ('actually') exist at any one time can only be finite. This is not the place for me to engage in arguments against that view.

As for the infinity of time, we need here to make a distinction. The *future* infinity is not a problem. Aristotle can happily accept there is an actually infinite sequence of days, one following the other, which stretches into the future without end. This sequence will never be completed. What seems to be more difficult for him is the *past* infinity. Since he holds that the world had no beginning, he must apparently accept that there are infinitely many days that have preceded this day, and that looks like a *completed* infinity. This is not a problem which he ever discusses, so I will merely make a suggestion on his behalf, one which stresses the importance of *the present*. We may suppose that his claim is that there cannot ever be a time at which an infinity of things—whether days or anything else—*does* exist. This is not to rule out the possibility of there being infinitely many of these things to come *later*, or infinitely many that have come *before*. For just as what will exist does not exist now, so also with what has existed. It is what *now* exists that must be finite, for any 'now', and this does *not* include the past.

Well, this—or something like it—*might* be what he would have said himself, but we can never know.

[1] In this case we must presumably suppose that the numbers are meant to be potentially infinite 'by addition', rather than 'by division'. Aristotle argues elsewhere that there could not be an actually infinite totality of the numbers, namely at *Metaph.* *M*. 8, 1083^b36–1084^a10, and *Phys.* III. 5, 204^b7–10.

8

A Note on Aristotle's Account of Place

For many years the scholarly consensus has been that there is something very wrong with the account of place that Aristotle gives us in *Physics* IV. 1–5. While there is some variation in the objections that different commentators have put forward, still a prevailing theme is that the account fails to yield the desired consequences for motion. This is because motion occurs when the same thing is now in one place and now in another, yet it seems that Aristotle's account makes it possible for things that are in motion nevertheless to stay in the same place, and for things not in motion nevertheless to change their place. Ben Morison has recently proposed an interpretation of Aristotle's account of place which defends it against these (and other) objections.[1] I wish to show that his defence fails.

Aristotle first defines the place of an object as the limit of the body that surrounds it (*τὸ πέρας τοῦ περιέχοντος σώματος*), and his original text may or may not have added that the relevant limit is that where the surrounding body touches the object (*καθ' ὃ συνάπτει τῷ περιεχομένῳ*, 212^{a}6–7). The text that we have inherited lacks this qualification, but in any case it must represent what Aristotle was thinking, even if he failed to say it.[2] A little later he reformulates this definition so that it adds the condition that the relevant surrounding body must itself be unmoved (*ἀκίνητον*, 212^{a}14–21), but one of the more interesting features of Morison's interpretation is his claim that this addition is not really needed. He concedes that the addition does represent an extension of Aristotle's initial thought, but nevertheless he claims that the first definition could have stood on its own, and would still have been entirely satisfactory.

To obtain this result he first argues that, when Aristotle speaks of 'the surrounding body', the word 'the' shows that he must be thinking of some *one* of the many different bodies that can equally be counted as bodies that surround our object *x*. (Several passages indicate that Aristotle is aware that one may consider either larger or smaller surrounding bodies, e.g. 209^{a}31–^{b}1, 211^{a}23–6, 212^{b}20–2). Which of them, then, is Aristotle referring to when he speaks of *the* surrounding body? Morison's interpretation is that he means the *maximum* surrounding body, for certainly Aristotle does think that there is such a body, namely the universe as a whole excluding *x*. (If *U* be the universe as a whole, then

[1] Morison (2002).

[2] The probability is that he did say it. See e.g. Ross (1936, ad loc.).

in mereological vocabulary this is $U-x$, i.e. the body that has as a part everything that does not overlap *x*.) But Morison does not, so far as I can see, offer anything by way of argument for this proposal (pp. 137–9).

An obvious alternative suggestion is that Aristotle is thinking not of the largest but of the smallest body that surrounds *x*. One might be inclined to object that there is no such smallest body, for if anything *y* surrounds *x* then so also does a suitably chosen proper part of *y*. However, there are indications within the discussion of place in *Physics* IV.1–5 that Aristotle might not agree. At $211^a29–^b1$ (cf. $212^b3–6$) he appears to be insisting that what counts as a body must be bounded by a genuine physical boundary, i.e. one that separates two physically distinct stuffs from one another. A purely notional boundary, drawn arbitrarily through the middle of a homogeneous expanse of the same stuff, he characterizes as one where what is on one side is 'continuous' (συνεχές) with what is on the other, and he does not allow this to count as a proper limit. (What he actually says, in the passage in question, is that such a boundary does not count as the *place* of what it surrounds, but it is very natural to add that this is because what it surrounds does not count as a body; it is only a part of a body, and a part of a body has no place of its own, but can only be said to be 'in' the whole of which it is a part.) Given this understanding of what is to count as a body, there will be both a least and a greatest body that surrounds *x*, and either might be what is meant by the apparently singular description '*the* body that surrounds *x*'. Moreover, if we have to choose between them, then I think the indications are that we should choose the least rather than the greatest. For Aristotle clearly thinks that the most basic place that a thing has is what he calls its 'proper place' (ἴδιος τόπος),which is no larger than it, and which he describes as what *first* surrounds the thing ($209^a31–^b2$, $210^b32–211^a3$, $^a27–9$). But this is a surrounding limit, and not a surrounding body. I do not think that he ever speaks of a first surrounding body,[3] but if he did it would surely mean the surrounding body that is closest to this surrounding limit, and this would have to be the least of the surrounding bodies, not the greatest. But I cannot pretend that this consideration is decisive, and it is more important that our text nowhere specifies *which* surrounding body is in question.

A natural explanation for this would be that Aristotle is thinking that it simply does not matter, for one will get the same result whatever surrounding body is chosen. (Thus the Greek article τό is being used not to signal a singular reference but to generalize.) This seems a plausible suggestion because, at least at first sight, we can say that whatever body surrounding *x* is chosen, the limit at which it is in contact with *x* will be exactly the same (for this limit will in all cases coincide with the limit of *x* itself). I am myself inclined to think that this is indeed Aristotle's (first) thought: he does not specify just which surrounding body is in question, because he considers the question to be irrelevant. Morison offers no

[3] But he should have done. See next note.

direct rebuttal of this suggestion, but it is quite clear what he would say if the point were put to him: often it *does* matter which surrounding body we choose, because it makes a difference to the question whether our object x is or is not moving. For example, it may be that y surrounds x and z surrounds y, so that both y and $y+z$ may be said to surround x. ($y+z$ is the mereological sum of y and z, i.e. the body that overlaps all and only those bodies that either y or z overlaps.) But then, if y is moving through z without changing its own shape in any way, then apparently the limit at which y meets x remains the same, while the limit at which $y+z$ meets x is changing. This is clearly the situation that Aristotle is thinking of when he has *second* thoughts at 212^a14–21. In the example, y is a vessel containing x, and this is a kind of 'movable place', because y itself is movable. But a place, says Aristotle, should be like an immovable vessel, and this leads him to his second definition: 'the place of x is the first unmoved limit of what surrounds x' (τό τοῦ περιέχοντος πέρας ἀκίνητον πρῶτον, 212^a20).[4] On this account Aristotle began by supposing that it would make no difference which surrounding body was chosen, but later realized that it must be specified that the chosen body is one that is itself unmoved.

These considerations are evidently not conclusive. I see nothing in the text which gives any support to the proposal that when Aristotle says '*the* body surrounding x' he means '*the maximum* body surrounding x', i.e. $U-x$. Certainly, he never says this. But at the same time he does not directly deny it, and we cannot be completely sure. It would be a good argument for the proposal if, as Morison claims, it is a proposal that rescues Aristotle's account from the objections that are commonly raised to it, so I now turn to explore this claim. For the sake of argument, then, I shall henceforth count as 'Aristotle's definition' what Morison takes him to mean, namely: the place of x is the limit of $U-x$ where it is in contact with x. Morison's leading thought is that Aristotle firmly believes that the Universe as a whole is unmoving. It follows that if there is any change in the geometrical configuration of $U-x$ this must be due to the motion of x and not of U. But also, it seems that we can argue that a change in what is defined to be the place of x, namely the limit of $U-x$, must at the same time be a change in the geometrical configuration of $U-x$. If so, then whenever x changes place (according to Aristotle's definition) then it also moves. And perhaps the converse can be argued too. That is the gist of Morison's proposal. I add that he therefore sees Aristotle's second definition of place not as introducing a new thought but as extending the old one. Since U is an unmoving body that contains x, we may also consider any other body y that is unmoving and contains x, and then $U-x$ and $y-x$ have the same limit with respect to x, so that if one alters then so does the other. Thus either limit may be taken to be the place of x, and it

[4] I am sure that what Aristotle meant to say was 'the place of x is the limit, where it meets x, of the first unmoved *body* that surrounds x'. (Here 'the first body' would mean 'the *least* body'.) But this is not what our text actually says, and what it does say seems to me to make no sense: how can *limits* surrounding x be distinguished as first, second, etc.?

will make no practical difference which is chosen. That is how the second definition extends the first, but without introducing any really new consideration.

Now, if to move is to change place, and a thing's place is a limit of some surrounding body, then the first question we must ask is: what is to count as a change of limit? Morison offers this answer: an ordinary object *x* (with one outer surface and no inner surfaces) counts as retaining the same limits if and only if (i) it remains the same object, and (ii) it remains the same size, and (iii) it remains the same shape (pp. 143–5). These conditions seem to me to be perfectly reasonable.[5] As Morison admits, we *also* have a tendency to think that the identity of a limit depends upon the identity of the matter of which it is a limit. His own example is that if one table-top is replaced by another, of the same size and shape, then we might wish to say that the table now has a different surface, and hence a different limit, even though the three conditions given are still satisfied (p. 144). In the other direction, if an ordinary rubber ball bounces on a hard flat surface then of course it is deformed while doing so. That is to say, it does not retain the same shape throughout, though we might well think that it does retain the same surface. Evidently Morison cannot accept such 'counter-examples' as these, for he wishes to say that an object may maintain the same place even though the matter in contact with it is changing (as when a boat is anchored in a river), and that an object may change place even though the matter in contact with it remains the same (as when the boat hoists its anchor and drifts down river).

Let us accept, then, that an ordinary object *x* retains the same (outer) limits just in case (i) it remains the same object, and (ii) it remains the same size, and (iii) it remains the same shape. It should be noted that on this account *x* may perfectly well retain the same limits while it moves from one place to another, or when it is turned upside down, and so on. (We may note, too, that Aristotle himself would surely accept this consequence, for his main argument against the suggestion that the place of *x* may be identified with the limit of *x* is precisely that when *x* moves it takes its own limit with it, but it does not take its former place with it; 209^{b}1–5 with ^{b}22–8, 211^{b}10–14.) What applies to the outer limit of *x* should therefore apply too to the inner limit of $U-x$, and it will remain the same so long as (i) $U-x$ remains the same object (and I shall assume that this condition is always satisfied[6]), and (ii) it remains the same size (which, we may assume, it will do if and only if *x* itself remains the same size), and finally (iii) it remains the same shape. Morison is therefore committed to maintaining that *x* changes its place (without changing its size) if and only if $U-x$ changes its shape.

He gives no general argument for this claim, but only an illustrative example, namely this: a disc with a hole at its centre is not the same shape as a disc with a hole

[5] I suspect that the first condition is not wanted. Suppose that Lot's wife is miraculously changed into a pillar of salt which retains the same shape and size as she had. The new pillar can hardly count as 'the same object' as the previous person, but might we not wish to say that it does still have the same limits? (This doubt is of no importance for what follows.)

[6] Alternatively, that it is not really required. See previous note.

near its edge, even though the two discs are the same size and the two holes are both the same size and the same shape (pp. 145–6). Similarly, then, if we think of the disc as representing Aristotle's spherical universe U, and the hole in the disc as representing an object x within it, and if x is at one time nearer to the centre of U and at another time further away from it, then $U-x$ has changed its shape from the one to the other. So $U-x$ has changed its (inner) limit, which on Aristotle's definition means that x has changed its place, and this result is exactly as it should be.

But we have only to reflect a little more on this example to see that it will not cover all cases. In particular it will not cover the case where x moves round the centre, keeping always an equal distance from it, and (if x is not itself a symmetrical object) keeping always the same orientation towards it. For in that case there is no reason whatever to say that $U-x$ has changed is shape. To see this clearly, we have only to revert to the homely example with which we began: if our disc starts with its hole above the centre, and ends with its hole below the centre, that is not a ground for supposing that the disc has changed its shape. For obviously we could transform the one situation into the other simply by rotating the disc, and we have already observed that rotating an object will not change its limits.

One might try to rebut the counter-example in this way: I described the situation as one in which x moves (round the centre) though there is no reason to suppose that $U-x$ has changed its (inner) limit. But perhaps we could insist on retaining Aristotle's definition of place, and so insist that, so long as $U-x$ does retain the same limit, then x simply has not moved. Bearing in mind the present example, would that yield a tenable position? The answer is 'no', as we may see by introducing a further object y, which also 'moves' round the centre, keeping always the same distance from it and (if relevant) the same orientation towards it. Then, just as $U-x$ has kept the same shape, so also has $U-y$. So, on Morison's criteria, there is no reason to suppose that either of them has changed its (inner) limit, and hence no reason to suppose that either x or y have moved. But I have not said anything which implies that x and y are moving in the same plane, or with the same angular velocity. So it may very well be that x and y are at one time close to one another and at another time far apart. Hence $U-(x+y)$ has certainly changed its shape, though $U-x$ has not, and $U-y$ has not. But we surely cannot accept that the composite body $x+y$ has moved, though neither of its two parts, x and y, have moved, so this solution will not do.

I remark, incidentally, that Aristotle's own universe contains an example of just this situation, namely in the motion of the planets (or of the spheres that carry them). On Aristotle's definition, as interpreted by Morison, it appears that each planet individually stays in the same place, yet they certainly move relative to one another. Is there some way of reforming Morison's account, so as to avoid this problem?

Aristotle constantly says that the universe as a whole is unmoving. It is true that his main argument for this claim is that there is no place beyond it for it to move to (e.g. 212^{a}31–^{b}3), and this implies only that it cannot move from one place to another, which would not prevent it from rotating in the same place. Moreover, he himself believes that its outer shell, i.e. the sphere of the fixed stars, does in fact rotate, even though on his account it has no place (ibid.). Consequently, I do not see that he has any strong reason to hold that the universe as a whole cannot rotate, but let us suppose—as is not implausible—that he could invent such a reason. This might then be used to throw some suspicion on the argument that I have been offering, for I made use of an analogy to a rotating disc. I said that the limit which separates a disc from a hole within it need not change just because the hole is at one time above the centre and at another below it, for we can bring about this change without changing either the shape or size of either the disc or the hole it contains, simply by rotating the whole disc. However, this analogy will halt if the universe, which the disc represents, cannot rotate, and perhaps we can say that the limit of $U-x$ must alter if x is sometimes above and sometimes below the centre of U, but U cannot move in any way at all, including rotation. This is not because $U-x$ alters in *shape*, for the same shape may occur in different positions. Rather, $U-x$ has changed its orientation, and it is intuitively obvious that if U itself cannot change its orientation then this can only come about through the motion of x. But, to rescue Aristotle's account, we need to explain why, in such a case, the limits of $U-x$ must have changed, and this is not intuitively obvious.

Perhaps Morison's criteria for when a limit remains the same could be modified so as to yield this result, though I do not see a simple way of doing so. (Recall that Aristotle must hold that a limit may remain unchanged even though the matter on either side of it is changing.) But, even supposing that this problem could be resolved, there would still be a difficulty remaining. For although we could now say that, since the planets move in relation to one another, there cannot be more than one planet x for which the (inner) limit of $U-x$ remains constant, there would still be no way of saying which planet—if any—this was. How would Aristotle himself react to this problem? Surely by invoking his firm (but unwarranted) belief that the earth as a whole is at rest at the centre of the universe, and does not rotate. So other portions of the universe will count as moving or unmoving according as they do or do not maintain the same distance and direction from particular landmarks on the surface of the earth. Given this criterion for being unmoving—a criterion which does not presuppose Aristotle's own notion of a place—then we could, if we wished, accept his *second* way of defining place: the place of an object x is the limit, where it meets x, of any unmoved body that surrounds x. But the first definition would simply have to be abandoned.

Obviously, this procedure is needlessly roundabout. For if we do at *some* point have to make use of the notion of a thing's distance and direction from some given objects, considered as unmoving, then we might just as well have defined a thing's place directly in these terms, and made no use of the notion of a *surrounding* body. This would allow us to set aside the unfortunate suggestion of Greek (and English) idiom, whereby we talk of a thing being *in* a place; for there is no good reason to suppose that a thing's place somehow 'surrounds' it. It would also open the way to thinking of a thing's place as 'relative', i.e. relative to certain objects which are being considered as unmoving, and this is surely as it should be. Morison hopes that he can obtain this relation to a thing considered as unmoving by invoking Aristotle's thought that the universe as a whole cannot move, but I have argued that it is not very likely that this was Aristotle's own (first) thought, and that anyway the idea does not work out. We obtain a more coherent account by invoking his belief that the earth as a whole is unmoving, but he himself never acknowledges the role that that thought is in fact playing in his assumption that places are 'absolute' and not 'relative'.

9

Aristotle's Account of Time

1. TIME AND MOVEMENT

Aristotle begins his positive account of time in *Physics Δ* by saying (218^{b}9) that time most appears to be a kind of change or movement, yet this cannot be quite right for two reasons: (i) that any change or movement has a particular location in space, whereas time does not, and (ii) that movements are fast or slow while time is not. Nevertheless, he goes on (218^{b}21–219^{a}10), time is 'not without movement', and in support of this claim he offers the argument that we notice that time has passed when and only when we notice that some movement has occurred. The premiss is false, and the argument is inadequate. The argument is inadequate because, of course, time may pass without our noticing it (as when we are asleep), and there is no obvious reason to think that during all that time there has been movement, even if we grant that when we do notice the passing of time that is because we notice some movement. But the premiss is false anyway, because we notice that time has passed not only when we notice that things *have* changed but also when we notice that they have *not*: that is, to be aware of rest, no less than to be aware of movement, is equally to be aware of the passing of time.[1] A useful example is this. Lying in bed at night one may be listening to the ticking of a clock, and be noticing no other change. Then one will—if one is giving the matter any attention—be aware that some time passes between each tick and the next. One may perhaps be aware that the *same* (or maybe a different) amount of time passes between each tick and the next. But one will also be aware that during those times between the ticks *nothing* (noticeable) happens. I do not think that Aristotle's doctrine can be rescued on this point, and I comment on it further at the end.

Having satisfied himself that time passes when and only when there is movement, Aristotle infers without any evident warrant that time must be a property of movement and shortly after concludes that in fact it is a 'number' (ἀριθμός) of movement. I think it fairly evident that this *must* mean *something* like 'time is a quantity of movement (viz. its duration)', but I shall look into this

Reprinted from *Phronesis*, 25 (1980), 148–69.

[1] At 221^{b}7–12 Aristotle himself remarks that time is the *measure* of rest as well as of movement.

very unexpected use of 'number' in the next section. For the moment, I just assume that the thesis is that time is some quantity of movement, in order to look further into Aristotle's account of the *temporal* nature of this quantity. It is, he says, a quantity of movement 'in respect of before and after' (κατὰ τὸ πρότερον καὶ ὕστερον). But what does this mean?

There is first the question whether this phrase is even supposed to be introducing a specification of the particular way in which time is a quantity of movement. Apart from the one sentence in which this definition is given (219^b1–2, repeated 220^a24–5), Aristotle usually speaks as if time is the *only* quantity of movement. But it is plain that this will not do. How much or how great a movement is *need* not be taken as a question about its duration, and in two places in Aristotle's own text this is evident. In 218^b15–16 he explains a fast movement as one where a thing 'moves much in little time', where obviously the *muchness* of the movement is explicitly distinguished from the amount of time it takes. Again, in 220^b26–32 he remarks that we also measure a movement by its *distance*, and a distance by the movement over it. One might further add that another relevant quantity is the *speed* of the movement, and of course in the seventeenth century there was a famous dispute as to whether the momentum or the kinetic energy should be accepted as giving 'the quantity of a motion'. If Aristotle's doctrine is to be acceptable, then, one must suppose that in his definition 'time is the number of motion with respect to before and after' the phrase 'with respect to before and after' *is* doing some work. In fact it must be supposed to be picking out the *temporal* aspect of motion as opposed to its other quantifiable aspects. Unfortunately it is not at all clear how it manages to do this.

Aristotle has begun by recalling his usual doctrine that every movement is from one terminus to another, and adding that the stretch covered by the movement is continuous, and so therefore is the movement, and so therefore is the time (219^a10–13). All these claims except the last are false for some movements (changes), but I do not propose to cavil about that now. Let us take it for granted that the kind of movement we primarily have in mind in this passage *is* the continuous movement of a body on a direct course from one position to another. Then, introducing the apparently important concepts of before and after, Aristotle says that these are notions which apply primarily in place, where they signify position, and that they are thence transferred to movement, and thence to time (219^a14–19). So we have apparently three different kinds of before-and-after to reckon with—in place, in movement, and in time. Furthermore, Aristotle uses the phrase 'the before and after' substantivally, as if it denoted an entity—or rather, entities—namely the things that are before and after. Can we say what they are?

I think we may first observe that these 'before-and-afters' are being construed as point-like and not stretch-like; for Aristotle says:

> We recognize time too when we distinguish (ὁρίζειν) movement, which we do by before and after. And it is then that we say that time has passed, when we perceive the before and

after in movement. We distinguish it [time? or movement?] by grasping that they [the before and after] are different, and that there is something else between them. For when we think that the extremes (ἄκρα) are different from the middle, and the soul says that the nows are two, one before and one after, then we say that there is time, and that that (viz. the middle) is time. For time is what is distinguished (ὁριζόμενον) by the now. ($219^{a}22$–30).

Now it is everywhere obvious that a 'now' is without duration for Aristotle. Further, it is clear from the end of the passage that we are thinking of two nows marking out the stretch of time between them, and it is these two nows which are characterized as before and after. Presumably, then, the same general picture will apply also to the movement, and to the space. In each case we shall have a stretch marked out by its two end-points, and it is the points which mark out (ὁρίζειν) this stretch which are referred to as the before and after.[2] If so, then the before and after in space are spatial points, just as the before and after in time are temporal instants. Further, a 'point' in a movement is evidently intended to correspond both to the spatial point where the body is at that time, and to the temporal instant when it is there, so I think we must construe it as the momentary state or event of the body being at that point at that time. The idea is that these momentary states or events are to be discerned in the first place by their relation to the spatial points they are correlated with, and that they may then be used in the explanation of temporal instants.

There is admittedly an objection to this interpretation, namely that Aristotle himself says that these 'points' in a movement are *movements*: 'the thing which at any time is the before and after in movement is a movement (though its being is something else and not movement)' ($219^{a}19$–21).[3] On our interpretation this commits him to the view that it is all right to speak of movement at an instant, which of course he later denies (*Phys.* Z. 3, $234^{a}24$–$^{b}9$, and often thereafter). I can only conclude that this is one of the rare occasions[4] when Aristotle falls in with the common way of speaking, which certainly permits us to say that a body is moving at this or that instant, despite his reasoned objections to it elsewhere. Anyway, since I see no satisfactory alternative to the interpretation suggested, I henceforth adopt it.

To return to our definition, we have apparently three interpretations to consider. Time is defined as a quantity of motion either (i) in respect of the before and after *in time* (i.e. in respect of temporal instants), or (ii) in respect of the before and after *in movement* (i.e. in respect of the momentary states of moving bodies), or finally (iii) in respect of the before and after *in place*. Now in the last case the quantity in question can only be the *distance* covered by the

[2] Admittedly in $220^{a}9$–11 ὁρίζειν is used differently, of *one* point dividing a stretch, and so marking out *two parts of the stretch* as before and after it.

[3] ἔστι δὲ τὸ πρότερον καὶ ὕστερον ἐν τῇ κινήσει ὅ μέν ποτε ὂν κίνησις· τὸ μέντοι εἶναι αὐτῷ ἕτερον καὶ οὐ κίνησις. I follow Ross's (1936) text and his interpretation of ὅ ποτε ὄν.

[4] The doctrine that there can be no movement at a moment is applied also to rest. But Aristotle carelessly allows himself to speak of rest at a moment at *Phys.* $236^{a}17$–18.

movement, so we can certainly rule out this alternative. It has seemed to many commentators (from Aquinas on) that the before and after in question must be taken to be the before and after *of movement*, just because to specify them as the before and after of time itself would seem to import an obvious circularity into the definition. But then the same objection as before seems to apply. Why should we take the kind of quantifiable stretch there is between two momentary states of a moving body to be a *temporal* stretch rather than once more a spatial stretch?

One can suggest a possible reply to this objection which certainly has some appeal. No doubt in the case of a spatial movement from one place to another the kind of stretch that there is between any two momentary states of it may perfectly well be taken as a spatial stretch. But Aristotle's definition is intended to compass *any* kind of change, as he later emphasizes (223^a29–$^b1$), and the stretch between any two momentary states of, for example, a colour change can hardly be regarded as spatial. Of course one might say that in this case it is a colour-stretch, and does not have to be a time-stretch, but now we can insist that the kind of stretch we are after is one that is found *both* between the momentary states of a change of place *and* between the momentary states of a change of colour, and indeed between the momentary states involved in *any* kind of change. Now on the interpretation I shall offer it will turn out that this is essentially Aristotle's reply, but clearly it is not the way he has presented the matter in our present passage. In that passage there is no emphasis on the many different kinds of change that time applies to, and the only kind of change under consideration seems to be a change from one spatial position to another. So for the present we must allow this objection to stand.

We must, then, say that time is defined as a certain quantity of movement, viz. that which is determined by the before and after *in time*, the temporal instants, or nows.* Indeed it has already been agreed (219^a30) that time is what is marked out—perhaps even defined (*ὁριζόμενον*)—by the now. If this is right, then either the notion of a temporal instant is to be taken as primitive (though illuminated somewhat by its comparison to a spatial point), or we have yet to complete the definition by saying more about instants. And Aristotle does have more to say about instants, for he seems to regard not only time, but also the now, as a *number*. We must, then, consider in more detail what Aristotle means by a 'number' in this connection. But I shall first ask how time is a number, and only later come back to the now.

2. TIME AND NUMBER

To his first and rather abrupt statement that time is the number of movement (with respect to before and after), Aristotle immediately adds three further

* See Additional Note 9.2.

elucidations. First, that time is not movement but that in respect of which movement *has* a number; next that time is a particular *kind* of number (ἀριθμός τις) and finally that 'number' can be understood in two ways, either as that which is numbered (or numerable) or as that by which we number, and time is number in the first of these senses (219^{b}2–9). All three remarks might naturally be interpreted as trying to make the same point, viz. that time is duration and duration is a kind of number. For first it is obvious that the relevant respect in which movement *has* a number is its duration, as is clear enough from the supporting point (σημεῖον) that Aristotle at once gives: 'we judge more and less by number, and we judge more and less movement by time' (^{b}4–5). From this he at once infers (ἄρα) that time is a kind of number (ἀριθμός τις), which harmonizes well with the suggestion that time is duration and duration a kind of number, and explicitly connects the second elucidation with the first. Finally we may add that that which is numbered—which is what time is—might naturally be taken to be the *duration* of the movement.

Now part of this interpretation seems to me correct: the first point that time is that in respect of which movement has a number, and the second point that time is a particular kind of number, are I think both intended to claim that duration is a sort of number. The claim does not consort very well with Aristotle's usual view that the numbers are just the natural numbers and are applicable to *discrete* quantities, whereas in the case of continuous quantities we speak of *measuring* and not numbering. But in our present discussion Aristotle is ignoring this distinction, or rather, is marking it in an unexpected way by the distinction between a number proper (ἀριθμὸς ἁπλῶς) and a particular kind of number (ἀριθμός τις). This emerges from the beginning of chapter 12, where we read:

> The least ἀριθμὸς ἁπλῶς is the number two (ἡ δυάς), but in the case of an ἀριθμός τις in a way there is and in a way there is not a least. Thus of a line there is least in plurality (πλῆθος), viz. two lines (or one line), but there is no least in magnitude (μέγεθος) . . . Similarly with time; there is a least in respect of number (κατ' ἀριθμόν), viz. one time or two, but there is no least in magnitude (κατὰ μέγεθος). (220^{a}27–32)

Evidently the fact there is no shortest line is here leading Aristotle to say that we have here a kind of number (ἀριθμός τις) for which there is no least. But what exactly is this kind of number?

It would be quite natural to *us* to think of these 'kinds' of numbers as what one might call 'applied numbers'. The idea would be that we can 'apply' numbers to lines in two ways: in one way—viz. when applied to measure plurality—there is a least number that can be applied, but in another way—when applied to measure length—there is no least. But I do not think this can be Aristotle's way of thinking of it, for this way presupposes the existence of some 'abstract' numbers waiting to be applied, and in the case of a magnitude these are 'abstract' numbers which have no least. They must therefore include the rational, and indeed the irrational, numbers. But it would be extremely remarkable if Aristotle is here assuming the existence of rational and irrational numbers as we understand them, since it would be quite without parallel in the whole of classical Greek mathematics.

I suggest, then, that we do better to think of this generalization of the concept of number in two stages. First we can think of a *natural* number as 'applied' to yield a number *of* something, and then we generalize from these applied natural numbers to something more like a 'quantity' or 'amount'. Thus when a natural number is applied to measure plurality we obtain such entities as are signified by '7 men' or '7 lines', and when it is applied to measure magnitude we obtain such entities as are signified by '7 minutes' or '7 inches'. These entities we may regard as *amounts*, respectively an amount of men, an amount of lines, an amount of minutes, and an amount of inches.[5] Now amounts of minutes or inches are special cases of amounts of duration and amounts of length, but not all such amounts are characterizable in this way as 'applied natural numbers', for we shall also want to include such amounts as 0.7 minutes or $\sqrt{2}$ inches. *We* refer to such amounts, as I have just done, by using an expression for a rational or irrational number in just the same way as we earlier used an expression for a natural number. But Aristotle, I imagine, would have proceeded more circuitously. The amount of duration that we call '0.7 minutes' he would perhaps characterize as 'the amount which bears to 1 minute the ratio of 7 to 10', and the amount of length that we call '$\sqrt{2}$ inches' he would no doubt refer to as 'the length of the diagonal of a square of side 1 inch'. He may well have recognized that there were amounts which he had no way of characterizing.

In this way one can generalize from the notion of an 'applied natural number' to that of an 'amount' without necessarily recognizing the existence of the rational and irrational numbers which could be 'applied' to yield these 'amounts'. So the interpretation I suggest is not altogether anachronistic, though I must confess it is a little surprising to find these amounts called 'numbers'. (The standard Greek expression for them is of course μεγέθη.) But what would be even more surprising is to find them called 'things numbered' as opposed to 'things we number with', for my point has been that the Greeks did *not* recognize any numbers with which you could number all these amounts. Rather, it seems to me that the amounts must themselves be the things we number with. But then, what do we number with them? What is the significance of the thesis that time is a thing numbered and not something we number with?

I believe some hint emerges if we look at the immediately succeeding sentence. As soon as he has made the point that time is a thing numbered and not something we number with Aristotle at once goes on: 'And as the movement is at

[5] We may compare the ἀριθμοὶ σωματικοί of *Metaph. N.* 5, 1092^{b}22. These are what are signified by the *terms* in a proportion such as '3 of fire to 2 of earth'. (But the units are left unspecified.)

every stage different, so also is the time (but all simultaneous time is the same, because its now is the same . . .)' (219^b9–11).[6] Now suppose that a time has just been characterized as an amount of duration. Then this sentence apparently continues by saying that the time of a movement, viz. its amount of duration, differs from moment to moment, whereas every simultaneous time, viz. amount of duration, is the same. But there is no proper opposition here, for clearly non-simultaneous amounts of duration may also be the same. And further, Aristotle's language strongly suggests that what makes the simultaneous times the same is something to do with just *one* now, but *one* now cannot pick out an amount of duration at all. Surely in this passage 'time' does not mean simply an amount of duration, e.g. 24 hours, but has something to do with the *date*. Is that the point of saying that time is a thing numbered?

Admittedly it is not obvious that these two sentences are supposed to be connected as I have suggested, but the connection is fairly clearly drawn in a later passage which apparently repeats our point:

> Time is the same at all places simultaneously, but earlier and later times are not the same, because also the present [stage of a] movement is just one, whereas the past and future [stages] are different [sc. from each other]. And time is a number, not by which we number, but rather as a thing numbered, and this is always different when earlier or later; for the nows are different. [Similarly] the number of 100 horses and of 100 men is the same number, but those of which it is the number—the horses and the men—are different. (220^b5–12)

It appears fairly clearly from this passage that the point about time being a thing numbered, rather than a thing we number with, is meant to ensure that earlier and later dates do not count as the same time. It may be remarked that in this passage, and in the passage cited earlier, Aristotle *appears* to slip between thinking of a time as a period and as an instant. This may possibly be so, and I shall later take up the question whether an instant may be referred to as a *time*. But anyway the point about 'the same time' implying the same date(s) is clearly applied to periods in chapter 14 at 223^b1–12. There the doctrine is that movements which begin simultaneously and end simultaneously both have the same time, though if they are equally long their times may be said to be 'the same in form' (*ὁ αὐτὸς χρόνος καὶ εἷς ὁ ἴσος καὶ ἅμα· εἴδει δὲ καὶ οἱ μὴ ἅμα*, 223^b3–4, Ross's text).

The full doctrine, then, seems to be this. There are such things as amounts of duration, e.g. 5 minutes, and the time of a movement is its amount of duration. But this is an ambiguous expression, depending on the extent to which the amount of duration of one movement is thought of as capable of being the *same* amount of duration as that of another. If we construe the expression quite universally, it simply applies to all movements which last equally long, no matter

[6] For the difficult continuation of this sentence I follow Ross's text and interpretation. But that does not concern us here.

when they occur. On this construal it would, I think, count as a 'number by which we number', though also of course as a 'particular kind of number' (ἀριθμός τις) rather than a number without qualification (ἀριθμὸς ἁπλῶς). Aristotle wishes, however, to construe it *less* universally, so that one and the same amount of duration is thought of as applying only to movements which are simultaneous with one another. And this, I think, is the point he is getting at when he says that time is a thing numbered rather than a thing we number with. But, so construed, the amount of duration *is* still a universal, not confined to this or that particular movement but shared by all simultaneous movements. It is partly for this reason, I believe, that Aristotle continues to call it a *number*. For when explaining his point that simultaneous movements do have exactly the same time he again makes use of the idea that time is a number. 'Of equal and simultaneous movements', he says, 'the *number* is one and the same, wherever they may occur' (223^b11–12), and once more he adds in comparison that 7 dogs and 7 horses have the same number.

One has to admit that the comparison is unfortunate and very confusing. For here Aristotle uses the point that 7 dogs and 7 horses have the same number in order to justify saying that simultaneous movements have the same time, as if the time answered to the number 7 'with which we number'. But earlier he has used the point that 100 horses is not the same thing as 100 men in order to justify saying that non-simultaneous occurrences do not have the same time, explicitly comparing the times to the different things numbered. On my account, the source of the trouble is that Aristotle has two quite different reasons for calling time a number, the first is that a time always has an amount of duration, i.e. it is a subject of which amounts of duration are predicated, and the second is that a time is itself predicated of the various movements that have that time. The second reason really amounts to no more than the point that a time is a universal, and is a very thin ground for regarding it as some kind of 'number'.

A possible escape from this confusion is suggested by the fact that my second reason is drawn wholly from a stray passage in chapter 14, and it is clear that the original discussion of time closes with the summary at the end of chapter 13 (222^b27–9). Chapter 14 then appears as a series of miscellaneous appendices added later, and is perhaps not to be reckoned as the work of Aristotle at all. I confess that I think this hypothesis is not easily overthrown, for there are other features of chapter 14 which strike one as suspicious.[7] I should have liked, therefore, to be able to draw upon more central passages to support my claim that *one* reason for calling a time a number is just that it is a universal. But what I shall do instead is to offer some indirect support, for I believe that this is Aristotle's *only* reason for calling the *now* a number.

[7] Miller (1974) points out that 222^b30–223^a15 is scarcely consistent with the account of 'being in time' given earlier at 221^a4–222^a9. Many commentators have looked askance at the suggestion of 223^a21–9 that time would not exist if there were no souls to do the numbering.

Before turning to this point, one more comment may be made here. If we do jettison chapter 14 we shall certainly not remove all inconsistencies in Aristotle's account of time, for there is one striking instance regarding the thesis that time is a thing numbered. At 220^{b}4–5 Aristotle is trying to explain why time is not said to be quick or slow, and says simply that no 'number with which we number' is quick or slow, evidently implying that time is a number with which we number, and in direct contradiction to the two places we have looked at in which time is said *not* to be that kind of number. There seems to be no satisfying explanation of this, except one that draws upon the *ambiguity* of 'time' between thing numbered and thing with which we number.[8] But one of the striking features of Aristotle's discussion of time is his *failure* to mention that the word 'time' has many senses. Some of these are well illustrated by the way the phrase 'a particular time' (χρόνος τις or χρόνος τις ὡρίσμενος) is quite ambiguous between a date pure and simple (yesterday noon), a dated temporal stretch (from noon yesterday to noon today), and a quantity of temporal stretch (24 hours). There are also other senses of this phrase. For example, when one says 'dinner is always at 7.30 sharp' one may properly be said to be giving a particular time as the time of dinner, but this time is what one might call a *recurring* date (and similarly with recurring periods). If Aristotle had paused to point out these ambiguities explicitly he would have saved himself from the appearance of outright contradiction on several occasions. (For example, 'earlier and later times are always different' (220^{b}9–10), and only three lines later 'one and the same time may occur again and again, e.g. a year' (220^{b}13–14).) In addition, his whole discussion of how time is measured by motion and motion by time (220^{b}14–221^{a}4; cf. 223^{b}12–23), which clearly has nothing to do with time as *date*, could have been much more clearly related to the passages we have just been looking at in which the time is the date. We are left to speculate on how far Aristotle was himself confused over these points—surely not as much might appear at first glance—but I do think it a defect of his discussion that he makes no attempt to lay them out for us explicitly. It is a defect which is all the more surprising, because it is very much in Aristotle's usual manner to do just this. But in our passage he never does it for 'time', but only for 'now' (222^{a}20–4).

3. THE NOW AND NUMBER

The hypothesis I aim to defend in this section is that Aristotle calls the now a number simply because it is a universal, viz. a date, holding of all the momentary events which have that date. Clearly he does view the now as a universal in this

[8] *Conceivably* it has struck Aristotle that time as a thing numbered *can* be said to be quick or slow. An amount of duration construed quite universally, e.g. an hour, cannot be said to be quick or slow, but we *do* say that this or that particular hour passed quickly or slowly.

way, and makes the point very early on (218^{a}25–7), but what is not so clear is that this is what he means by calling the now a number. One might suggest that Aristotle was influenced here by the fact that we do often use numerals to refer to dates or instants—numerals which refer primarily to the *amount* of time between the date in question and some fixed date. But in fact I do not think this at all likely, for there cannot have been many precise instants which were thus referred to in Aristotle's day, owing to lack of appropriate clocks, and there is no hint of this motive in anything he actually says. He nowhere suggests any reason for associating particular nows with any of the more familiar numbers.

I think we can go further. It has been suggested that a now is a number because we use nows to *count off* a period of time. As Callahan (1948) puts it:

> Suppose I begin reckoning time from the present now. At certain intervals I count one minute, two minutes, three minutes, and so on. The lapse of an interval is denoted by an indivisible moment, the present now. When three such nows have been counted, as reckoned from the original now, we may say that the motion which is being reckoned can be represented or numbered consecutively by three units, each inextended and discrete . . . (p. 52)

Again, a little later

> We recognise motion by perceiving the moving object at different phases, and the time by perceiving different nows. So the now numbers the moving object by corresponding to these phases in the order in which they present themselves to the mind. The sum total of nows constitutes time as number by being the sum total of phases as distinguished by the mind in the continuity of motion. (p. 58)

But this interpretation surely conflicts very noticeably with Aristotle's firm grasp of the point that time is continuous and nows have no duration. He often insists that a period of time cannot be made up of nows, and he never once hints that only certain nows are numbers, e.g. the ones that a mind can distinguish, or ones that are equally spaced along the continuum of time. On the contrary, it appears that *every* now is a number, and it is of course a well-marked feature of his doctrine that there are no smallest periods of time, and that all periods however small—including, evidently, those periods which are too small for us to notice (222^{b}15)—are bounded by nows and contain further nows within them. These are points so firmly established in his thinking that it is incredible that he should here overlook them. It cannot be, then, that nows are numbers because of some association with the *natural* numbers.

More generally, Aristotle must surely have seen, from the connections that he is constantly drawing between space, time, and motion, that just as there are incommensurable lengths so too there must be incommensurable periods of time, and hence if all nows are to be associated with numbers the numbers in question must include irrational numbers. Now we have seen that in a sense Aristotle does allow for irrational numbers, viz. as what we may call 'irrational amounts', but obviously a now is not an amount. *Two* nows may be said to mark out an amount, but that does not entitle either of them singly to be called an amount—or, if it does, then we *must* be given some explanation of how. But our text contains no such explanation. We can conclude, then, that nows are not numbers by being associated in some way with more familiar numbers. When nows are called numbers it must be because they have some more distant analogy with the familiar numbers. My suggestion is that the point is simply that nows are like numbers in being universals: a now collects together all the various momentary events which occur at it, and this Aristotle puts by saying that the now 'numbers' those events. But no more than this is intended.

It is now time to see how this interpretation is borne out by our text, and as there are three passages in which Aristotle connects the now with number I shall discuss them in turn. But first we must notice a complication which bedevils this problem, viz. that Aristotle is here deliberately using 'now' in two distinct senses. His standard practice, of course, is to use the expression 'a now' simply to mean a temporal instant or moment, without any real connotation of presentness. But here he wishes to maintain that though in a way each now is different from every other, in another way they are all the same, just because in a sense there is only one present. To make this point he compares the now to a body in motion, saying that we have the same body throughout the motion, though it can also be said to be always different in so far as it is first a body-in-this-place and later a body-in-that-place. Similarly we can regard the now as always one thing, while at the same time it can be said to be always different, presumably because it is first a now-of-this-event and later a now-of-that-event, the events in question being the different momentary stages of a single movement, i.e. the things that are 'before and after' in that movement. As Aristotle puts the matter, the *thing* which at any time is now (τὸ νῦν ὅ ποτε ὄν) is always the same, but what it is for it to *be now* (τὸ εἶναι αὐτῷ) is always different, presumably because that is what makes it *now* rather than *then*. On my interpretation, it is the now which is always different which is a number, and not the now which is always the same. I confess that the text is not always as clear on this as one could wish, but at least our first passage suits this part of the interpretation.

Aristotle is in the course of drawing out his comparison between the now and the moving body, and says

> The now answers to the body in motion, just as the time answers to the movement, and it is in so far as the before and after is numerable that it is (or: there is) the now. (219^{b}22–5)[9]

The train of thought is very condensed, and there is no explanation of what it might mean to say that the before and after is numerable, nor how this connects with the now. But it can be seen that the connection must be with the now which is always different if we look at the three lines following when the same claim is repeated. For Aristotle at once continues that in this case too (καὶ ἐν

[9] . . . ᾗ δ' ἀριθμητὸν τὸ πρότερον καὶ ὕστερον, τὸ νῦν ἔστιν.

τούτοις—viz. in the case of time and the now, as previously in the case of movement and the moving body)

> The thing which at any time is now is (always) the same, since it is (always) what is before and after in movement, but its being is (always) different, because it is in so far as the before and after is numerable, that it is (or: there is) the now. (219^b26–8)[10]

Here it is explicitly the *difference* from moment to moment that is connected with the *numerability* of the before and after. Indeed this latter passage would permit the view that to say that before and after is *numerable* is no more than to say that it is *distinguishable*—that distinct (momentary) stages can be discerned within it. (So Ross.) But I think this rather simple interpretation is inadequate to the passage we began with, so let us now return to it.

Aristotle is developing or explaining his claim that the now answers to the body in motion just as the time answers to the movement, and the analogy he has in mind seems to be this. We recognize earlier and later stages of the movement by recognizing that the moving body is first in this place and then in that. But time is that in respect of which the movement is numerable, and the now is that in respect of which the stage of the movement is numerable, so the whole situation will transfer from movement to time. Hence we shall similarly recognize earlier and later stages of the time, i.e. earlier and later nows, by recognizing that the one persisting now is first in one state and then in another. If this is right, then the source of the parallel is that the now is the number of the stage of the movement just as time is the number of the movement as a whole. Each stage separately, then, has a now as its number, so it will not do to explain the numerability of the stages as no more than the fact that they are distinguishable from one another. Rather, Aristotle is adverting to the relation between a now and a movement-stage, comparing it to the relation between a time and a movement, and in each case describing the relation by saying that the one is a number of the other. I therefore interpret him thus. The reason why the movement is 'numerable' and its time is its 'number' is just that it has an amount of duration which it shares with all other simultaneous movements. So too each particular stage of the movement is 'numerable', and its now is its 'number', just because it has a date which it shares with all other simultaneous events. Let us now turn to the next passage.

In order to emphasize that time and the now depend upon one another for their existence, Aristotle apparently returns once more to his comparison, and says

> Just as the body in motion and the movement exist simultaneously, so also does the number of the moving body and the number of the movement. For time is the number of the movement, and the now is as the body in motion, like a unit of number. (220^a1–4)

[10] ὅ μέν ποτε ὂν νῦν ἐστι, τὸ αὐτό (τὸ πρότερον γὰρ καὶ ὕστερον ἐστι τὸ ἐν κινήσει), τὸ δ' εἶναι ἕτερον (ᾗ ἀριθμητὸν γὰρ τὸ πρότερον καὶ ὕστερον, τὸ νῦν ἔστιν).

Now at first glance it seems that it must here be the *persisting* now which is said to be the number of the moving body, first because it is said to exist simultaneously with the number of the whole movement—i.e. throughout all the time of the movement, apparently—and second because it is compared to a *unit of number* (οἷον μονὰς ἀριθμοῦ). On the face of it, a *unit* of time would have to be a *stretch* of time, with longer stretches being measured by taking a number of such units.[11] But on reflection one sees that these two reasons are in conflict with one another. The comparison introduced by 'like a unit of number' must surely be equating the now to the unit *in contrast* to the time of the movement as a whole which is the relevant number. But it would be absurd to bring in the distinction implied between unit and number if what Aristotle intends to do is to compare the now that persists through the whole movement with the unit, and the time of the whole movement with the number. For, as our first reason maintains, the now that persists through the whole movement does literally exist *simultaneously* with the time of the whole movement, and they must in fact be exactly the same thing. The 'flow' of 'the present' from the beginning of the movement to the end just *is* the time of the movement, just as the 'flow' of the moving body simply is the movement.

I think, then, that this interpretation breaks down, and we must suppose that the now here in question is once more the momentary now. In that case its characterization as 'the number of the moving body' must be treated as shorthand. Just as Aristotle said earlier 'it is by means of the body in motion that we recognize the before and after in movement' but obviously *meant* by this that we recognize the before and after by recognizing that the *same* body is at *different* positions (219^b23–5; cf. 219^b17–18, 220^a9), so here the phrase 'the number of the moving body' must be taken as short for 'the number of the moving body's being at this or that position'. The claim that this exists 'at the same time as' (ἅμα) the number of the movement as a whole must then be regarded as a somewhat loose expression. Perhaps indeed ἅμα is being used here without any temporal connotation, but simply as a graphic way of saying that neither can exist without the other. Finally, the comparison with a unit of number, which is indeed 'unfortunate' as Ross says, must again be explained as intended simply to emphasize the mutual dependence. The idea is that just as there would be no (natural) number of anything if there were not one of it, so equally there would be no stretch of time (or any other such magnitude) if there were not points of it. The comparison, then, is not very exact, but is perhaps good enough for the present purpose. I conclude that we can explain this passage on the assumption that it is the momentary now which is in question, and hence the interpretation I first suggested is still available. But this passage sheds no more light than the last on what it *means* to say that a now is the

[11] Elsewhere Aristotle takes one revolution of the fixed stars to be the unit (μέτρον) of time (*Metaph. I.* 1, 1052^b35–1053^a12).

'number' of an event which occurs at it. Let us move on, then, to our next and final passage.

The last place where the now is called a number is 220ᵃ21–4, but we shall need to observe first how Aristotle leads up to it. He has remarked that the now is what makes time continuous, and equally what divides it. His point seems to be that whereas the momentary now divides time into its various portions, it is the persisting now that makes time a continuum, just as it is the fact that the moving body persists as one and the same body which makes a movement one, and similarly too for the spatial point which generates a continuous line by 'flowing' along it.[12] At any rate, that is how the passage is recapitulated at 222ᵃ10–17, though if we take into account Aristotle's definition of what it is to be a continuum—viz. to divide into parts which touch each other at the same boundary (*Phys. E.* 3)—the point about 'flowing' hardly seems necessary. But anyway it is clearly the momentary now which divides time, even if it is the persisting now which holds it together (συνέχει) so that it is continuous (συνεχές). There then follows a rather telescoped remark about *time* being a number, viz.

Time is a number not as [a number] of the same point—on the ground that it is both a start and a finish—but as the ends of the line[13] [are a number?], and not as the parts [are a number?] (220ᵃ14–17)

It may not be quite clear why the fact that a dividing point is both the start of one stretch and the finish of another might have led us to suppose that it had a time for its number, but it seems that the explanation is this. Aristotle thinks that when we treat a spatial dividing point as both start and finish there must be a pause at the point (ᵃ12–13), and perhaps he is adverting here to his view that points do not really exist until they are *actualized* in some way, and this always takes time. (Cf. *Phys. Θ.* 8, esp. 262ᵃ12–ᵇ8, 263ᵃ23–ᵇ9.) So if the same were to hold for the now, considered as dividing point, the now would have to occupy a stretch of time, and so have a time for its number. But in fact the now has no duration. (So, apparently, the now *does* divide time, but only potentially and never actually, for we cannot hold it still long enough to do any actual dividing with it. Cf. 222ᵃ12–14, 17–20.) Anyway, the now is not a part—i.e. a stretch—of time, and time is a number by being a stretch marked out between two nows.[14]

[12] For the general doctrine cf. *De Anima A.* 4, 409ᵃ4–5. That the 'flowing' point may be regarded as the *same* point at its various positions on the line is explicitly stated at 222ᵃ17, and I suspect it is also the meaning of 219ᵇ19. (Should we perhaps read: ἡ στιγμὴ γάρ, ἢ [ὁ] λίθος ἢ...?).

[13] Ross's suggestion that τῆς γραμμῆς be omitted seems very likely right. But the sense would be unaffected.

[14] The suggestion that time is a number 'as the parts' seems not to be pursued. The two reasons given (ᵃ17–20) for endorsing the view that time is a number 'as the ends of a line' seem both to be reasons for denying that it is a number 'as of the same point'. But perhaps Aristotle's thought is this: since the parts of a time are themselves times, and therefore numbers (cf. αἱ δὲ γραμμαὶ αἱ δύο τῆς μιᾶς μόρια, ᵃ20–1), it would be circular to invoke them to explain how time is a number.

After this comes our passage, which says:

In so far as the now is a limit it is not time, but belongs to it. But in so far as it numbers, it is a number.[15] For limits pertain only to that of which they are limits, whereas the number of these horses—ten—may hold elsewhere too. (220^a21–4)

It is clear that the now as a limit is the same thing as the now as a dividing point, the start of one part of time and the finish of another (cf. 222^a11–12). Concerning this, we have just said that it is not what makes time a number, since it has no duration, and hence we are told here that it is not (a) time. But, the sequence of thought seems to be, although not a number in this way (viz. by being a time, which is a number), it still is a number in a different way, i.e. 'in so far as it numbers'. That at any rate is what the manuscript reading suggests, and it seems just possible, though certainly very tortuous. Besides, the statement that 'in so far as it numbers it is a number' is uncomfortably vacuous.

It is therefore worth considering the alternative reading found in Philoponus,[16] which omits 'it is a number' and consequently understands 'it is time' instead. This gives a very straightforward antithesis, viz. that what makes a now something *temporal* is not that it is a limit but that it numbers, and it makes very good sense of the whole passage. But I must admit that it is a serious objection that Aristotle simply does not count the now as 'a time', since all times must, for him, have duration. It is, of course, perfectly ordinary Greek usage to call a particular instant 'a time', and one that Aristotle falls in with elsewhere (e.g. *Cat.* 5^a7), but here he is very consistent in avoiding it. (The only strong candidate for an exception in these five chapters of the *Physics* is 219^b10 (above, pp. 140–1), and even that seems to me uncertain.) I shall not attempt to decide between these two suggestions.

What is important for our purposes is that we have here a contrast between the limiting function of the now and its numbering function, and then the passage goes on to explain the relevant difference between them: 'limits pertain only to that of which they are limits, whereas the number of these horses—ten—may hold elsewhere too'. Clearly the crucial feature of a number is just that it is universal. It is this which entitles us to call the now a number (and *perhaps* also what entitles us to regard it as a time). The boundary of a particular movement is not a number, since it is the boundary merely of that movement and not of anything else. The boundary even of a stretch of time is not a number simply by being a boundary. What makes it a number is that the boundary of a stretch of time is *also* the boundary of all movements that have that time, i.e. it is a universal. So far as one can see, no more than that is needed.

The defence is concluded.

[15] ᾗ δ' ἀριθμεῖ, ἀριθμός. This is the reading of all our MSS, but it appears that Philoponus lacked the word ἀριθμός and understood χρόνος ἐστι instead. (See Ross 1936, ad loc). I discuss this alternative reading below. (Ross obelizes the passage.)

[16] See previous note.

4. EVALUATION

We have now looked at what I take to be the essential features of Aristotle's account, and before embarking on a critical appraisal it may be well to summarize them.

The basic notion is that of the now, and it is the momentary now—the now simply as temporal instant—that bears the weight of the account. For it is this that divides and bounds the parts of time, and time itself is defined as the quantifiable stretch that there is between any two such instants. An instant is to be construed as a common property of various momentary states of bodies, namely the momentary states that divide and bound their movements. And these in turn are to be picked out by the spatial points which the bodies occupy during their movements—points which themselves divide and bound the spatial stretches over which the movements occur. This is how we first grasp the notion of a momentary state: it is the state or event of a body's being at this or that position in its movement. But the notion is then to be generalized—Aristotle does not really explain how—so that we can discern momentary states within any kind of change, and not just a simple movement from one position to another. Finally, two momentary states share the *same* instant if and only if they are simultaneous, and this, the concept of simultaneity, is the primitive temporal concept that is used and not further explained. A temporal instant, then, is the property which all members of a set of simultaneous momentary states have in common, simply by virtue of their simultaneity.

If we had taken Aristotle's alternative account of an instant, viz. as a particular determination—in fact a momentary state—of that 'now' which travels always with us, then we should be brought back to what at first sight seems essentially the same position. For it would be important to explain that the different momentary states of this persisting present are themselves distinguished by its being now *the present of* one momentary state of an object and now *the present of* another. But then we should have to add that all *simultaneous* momentary states of objects determine the *same* momentary state of the present, and so once again an instant is determined by a set of simultaneous momentary states of objects. But on this alternative approach the underlying ontology appears rather different. On the first approach an instant is a property (viz. its date) of a momentary state of an ordinary physical object, while on the second approach an instant is itself a momentary state of an alleged persisting entity that I find rather obscure, despite Aristotle's assurance that it is 'most knowable' (219^b29). He compares it, indeed, to a substance (a 'this'), but it is certainly not any ordinary kind of substance. I think, then, that this alternative approach is rather threatening to the ontology that Aristotle ultimately desires, and so I propose not to take it any further. (On *one* reading of my last text on the now (220^a21–4, p. 149 above) I have Aristotle's own support: an instant is not a time by being a limit of the movement of the present—a momentary state of the present—but because it *numbers*, i.e. it is predicated universally of many things.)

Reverting to the first approach, let us call any momentary state of an object an event, so that instants are properties of events. These properties Aristotle regards as *numbers*, though it must be admitted that they are numbers only in a most tenuous sense, viz. by being properties corresponding to equivalence classes of events generated by the equivalence relation of simultaneity. Correspondingly, a time—that is, a stretch between two such instants—is the common property of all processes which are bounded by events belonging to those instants, or in other words it is again a property corresponding to an equivalence class of processes generated by the relation of simultaneity between processes. This is *more* appropriately called a 'number', because at least it *has* a 'number' in the extended sense by which an amount of duration counts as a 'number'. The amount of duration that a time has is (in suitable cases) ascertained by measurement, which again is primarily and in the first place a measurement of the processes that have that time.

Such, then, is Aristotle's basic scheme for explaining time. It is easily seen to be inadequate, for it is obviously impossible to explain all aspects of time by starting with the concept of simultaneity as one's only primitive temporal concept. At the very least we must add the concepts of earlier and later, which certainly cannot be derived from the 'before and after' in place, as Aristotle seems to imply.[17] We should also do well to include the concept of equal duration as a further temporal primitive, for—briefly—the *order* of time does not determine its *metric*. (Aristotle's own discussion of the measurement of time, in 220^{b}14–24, 221^{a}1–4, 223^{b}12–18, evidently presupposes and does not explain the required notion of equality.) Still, the fact that Aristotle's account needs to be expanded in these ways does not disturb what one naturally takes to be *the* main claim in Aristotle's account of time, viz. that concerning its ontological status. The extra temporal concepts that are required can clearly be viewed as concepts which apply primarily to movements and their limits, and so we can accept them as irreducible without abandoning Aristotle's fundamental doctrine that time is ontologically dependent on movement (or rather, change), and that that in turn is ontologically dependent on substances. Of course the second part of this claim is not very secure: for example, it is not at all clear that the event reported by 'there was a flash of lightning at noon', or the process reported by 'it then rained all afternoon', are to be regarded as the momentary state, or the motion, of any *substances*. But I leave that line of criticism aside, for it is the first part of the claim that relates most directly to the nature of time. Let us grant, for the sake of argument, that every event or process can be viewed as an affection of one or more substances. Should we grant also that time is essentially a property of these processes and events?

[17] The point is well argued in Owen (1976: 23–5).

I pointed out early on (p. 136) that Aristotle's argument for connecting time and movement fails, for we can perceive rest as well as movement. The point may be strengthened. There is no conceptual difficulty in the view that time may pass while nothing at all happens, for there is no conceptual difficulty in supposing that the universe consists *just* of a ticking clock, whose ticks are separated by time, and are caused not by any *continuous* process in the clock but in some mysterious and *intermittent* manner (or, perhaps, are simply not caused at all). Indeed we have no guarantee that our own universe is not essentially like that. One could roughly imagine a development of quantum theory which would posit the existence of some very short time intervals—let us call them chronons—such that the whole state of the universe remains unaltered throughout each chronon and then 'jumps' discontinuously to a new state at the instant separating that chronon from the next. Let us grant, then, that the passing of time does not strictly require any movement or change.[18] Is this a fatal blow to the Aristotelian position?

So far as I can see, it is not. Admittedly, time passes both when there is motion and when there is rest, and we concede that rest may occur when motion does not. But if we are granting to Aristotle, at least for the sake of argument, that when there is motion it is always the motion *of some substance*, then should we not equally grant that when there is rest that too is because *some substance* is at rest? And if we did grant this, would it not be enough?

For lack of space, I propose to grant without further discussion that when there is rest that is because something, and we may presume some substance, is at rest. I shall also grant that time passes only when there is either movement or rest. But still it may seem that this is not enough to rescue Aristotle's position. For if time may pass while all is at rest, how is Aristotle to introduce the notion of a momentary state, which is fundamental to his account? As we saw, his own idea was to say that the different momentary states of a body in motion are the states of the body being at this or that different spatial position. Clearly this presupposes that the motion is uninterrupted (and, incidentally, that it does not repeat itself). But at best this explains momentary states only for bodies in uninterrupted motion. From here it might seem that we can extend the notion to bodies at rest, but only *provided* that there always is some uninterrupted motion going on somewhere. For then we can say that a momentary state of a body at rest is a state of the body that is simultaneous with some momentary state of a body in uninterrupted motion. But clearly in a wholly static universe, or even in our universe with chronons, that explanation is not available, for in these universes there is *never* any uninterrupted motion.

A slightly different approach to a momentary state or event, but still clearly Aristotelian, would at least yield such events for the point of transition from one chronon to the next. We simply construe a momentary event as the boundary

[18] See Shoemaker (1969) for a different argument to the same conclusion.

that divides one more or less lasting state of a body from another, e.g. the boundary between a body's being to the left of a certain point and its not being to the left of it. But at first sight it appears that we shall not be able to distinguish different momentary states within the same chronon. In consequence, there will be no instants within a chronon either (since an instant exists only when there is a momentary event occurring at that instant), and the structure of time, i.e. the order and arrangement of the instants of time, will be discrete. This is not to say that there will be no time-*interval* between the two instants which bound a chronon, but just that this is an interval which contains no instants within it. I can imagine that Aristotle might be perfectly satisfied with this conclusion: the structure of time, he might say, is indeed dependent on the things that happen in time.

If the conclusion seems unsatisfactory, it is because we can certainly divide the time-interval in our thought. We can say, for example, that things are just the same at the beginning of the chronon as they are *halfway through it*, and furthermore this remark is by hypothesis true. How could that be, if there was no such thing as being halfway through a chronon? But the objection itself shows the way to meet it. For if we can help ourselves to another temporal concept, that of equality of duration, then clearly we can use this to define new momentary states in terms of their temporal relations to antecedently given momentary states. As earlier we relied on the concept of simultaneity to transfer the notion of a momentary state from a continuously moving body to one at rest, so now we rely on equal durations in much the same way. In fact we need only be given two momentary states to define all others in this way.

I confess that I cannot see how to do this trick with only *one* momentary state antecedently given, and certainly not with a completely static universe in which there are none. But does this matter? The difficulty we have been discussing is that on Aristotle's approach we must make good an *ontological* claim, that an instant exists only when there is a momentary event that occurs at that instant, and a claim about our *understanding*, which I have represented as the insistence that we must explain what an instant is in terms of what a momentary event is. The ontological claim can be upheld even for the wholly static universe: we may insist that if there is time in that universe then there are instants in it, and there are also momentary states. The fact that we are given no way of picking out any of these momentary states surely does not prevent us from claiming that they are there. As for the point about explanation, it is of course true that if we were actually faced with a wholly static universe, and could not even envisage other kinds of universe to use as examples, then we could not explain the notion of an instant by first explaining the notion of a momentary state. But then there would be no *other* way of explaining what an instant is either, and indeed no possibility of our understanding either that notion or any other temporal notion. And this is surely the fundamental truth that lies behind Aristotle's thesis that time depends on motion.

Additional Note 9.1. General

This essay addresses two main questions of interpretation: (i) why should Aristotle call time 'a number', and (ii)—even more surprising—why should he call the now 'a number'? It seems to me that there has been much less astonishment over these claims than there should have been. The essay here reprinted offers answers, but I confess that I cannot still believe these answers. However, I know of nothing better to believe, and I do think that the claims are very puzzling. I here elaborate on this point.

Aristotle thinks of time in much the same way as we do (in our ordinary and unrelativistic thought), i.e. as a single linear continuum, the same for all objects and events at all places. In this way it resembles a spatial line, thought of (no doubt) as a straight line, but in any case as one that nowhere intersects itself. There is a difference which was important to Aristotle. According to him the linear continuum that is time has no beginning and no end, whereas every straight line will have both, for there can be no straight line that is longer than the (finite) diameter of the universe.[19] But I think that this difference is of no real importance to us, so for the purpose of this discussion let us remove it, by thinking of space as infinitely extended in all directions. Then in all important topological respects time resembles an infinite straight line in space.

How would you understand someone who said 'Space is a number'? Or, perhaps a little more comprehensibly, 'A spatial distance is a number'? And what if he added 'and every point on that distance is also a number'? What *could* he mean? Well, the natural response from our point of view would be to begin from the thought that we do use numbers to describe how long a distance is. But here a crucial difference is that we are familiar with the notions of a rational number and an irrational number, whereas Aristotle presumably was not (for these notions do not figure in classical Greek mathematics). Suppose, then, that we are looking at a square table-top, with sides each 1 metre long. (And assume, for simplicity, that this top is a *perfect* square.) Now consider someone who says 'There is a distance between the opposite corners of this square, and (like every other distance) it is a number'. We presume that he must mean that the length of this distance is some definite number of metres (or other units), and when we ask him '*Which* number?' we expect him to reply '$\sqrt{2}$ metres' (or some equivalent). If we ask him further about which number this is, we might expect him to say something like 'Well, it is a little larger than 1.4 metres, but considerably smaller than 1.5 metres; in fact $\sqrt{2}$ is the unique real number that is greater than every rational number whose square is less than 2, and smaller than every rational number whose square is greater than 2'. *We* can accept that kind of answer because *we* have been brought up to believe that there is such a number. But

[19] Cf. *Phys.* III. 7, 207^{b}5, 15–21, 27–34.

Aristotle was not. *He* was brought up to believe that there is *no* number which, when multiplied by itself, yields 2. So how could he think that the answer '$\sqrt{2}$ metres' mentions a *number*? And if that is not the answer that he has in mind, then what would he think of as an appropriate answer to the question 'And which number is that?'

One might suggest that in our way of thinking a ratio, such as the ratio of 3 to 4, can regarded as a number, namely $\frac{3}{4}$. In classical Greek mathematics a ratio is not itself taken to be a number, but still ratios are akin to numbers. For every ratio can be described by its relation (larger or smaller) to the ratios between natural numbers. So perhaps we could stretch a point and allow that a ratio *is* a number. But still, a ratio is a ratio *between two* magnitudes (distances, or times, or weights, or whatever). Would it make any sense to say that a distance *is* a ratio? That seems to me very improbable. (Certainly Aristotle would believe that any two finite distances do stand to one another in some definite length-ratio, for Eudoxus had explained how to make sense of this, and Aristotle was familiar with the explanation.[20]) But still, a ratio is a relation that holds between two distances (or other magnitudes), and is not a property of one distance by itself. So how would that entitle us to say that a distance *is* a number? Yet, if we cannot explain in this or any other way what it might mean to call a distance 'a number', how could it make any better sense to say the same of a time?

Annas (1975) makes this suggestion:[21] Aristotle's reason for calling time 'a number' was just that he was already firmly convinced (against Plato) that numbers had no independent existence.[22] So this seemed to him a natural way of saying that the same applies to time, i.e. that time has no existence of its own over and above the existence of the various changes that take place in time. No doubt this is something that he does wish to say, but one still cannot see why it leads him to call time 'a number'. After all, he also believes (against Plato) that the form of man has no existence of its own over and above its exemplification in particular men, but that does not lead him to say that this form is 'a number'. Besides, he was obviously well aware that he and Plato took very different views over the 'separate' existence both of numbers and of the form of man, but in fact it is not at all clear that they did differ over the 'separate' existence of time. The view that time depends upon change—in particular upon the regular rotation of the heavens—seems common to both of them.

Commentators naturally think of Newton's distinction between 'absolute, true, and mathematical time', which 'of itself and from its own nature flows equably without relation to anything external', and 'relative time', which is 'some

[20] Cf. e.g. *Posterior Analytics* I. 5, 74^{a}17–25.

[21] I regret that, in order to prune my original essay to an acceptable length for publication, I had to remove all references to Annas (1975).

[22] Annas refers in particular to chapters 1–3 of book *I* of the *Metaphysics*, but I myself do not see the view in question expressed more clearly there than in several other places where hints are dropped. But in any case no one doubts that this *is* Aristotle's view.

sensible and external measure of duration by the means of motion, which is commonly used instead of true time' (Newton, *Principia*, Scholium I). (For example, the distinction is twice quoted by White 1992: 75 and 84, who very fairly says that Aristotle accepts only the idea of 'relative time'. Cf. also Sorabji 1983: 79–80.) Annas, in her (1975: 105–6), describes a view of time which is certainly very similar to Newton's 'absolute time', and which she regards as a 'Platonist' view that Aristotle was meaning to attack. But Plato himself did not hold such a view, and I know of no reason to credit it to anyone whom Aristotle would have been familiar with. It is not obvious that he recognized that he had any opponents on this topic.

I cannot, therefore, come up with any good motive for Aristotle's strange claim that time is 'a number', and I wish that there had been more discussion by others of what this could mean. It is usually discussed in too simple a way (i) by observing that *later* in his discussion he quite frequently calls time 'a measure' rather than 'a number', though in fact measures and numbers are certainly not to be identified in standard Greek thought;[23] and (ii) by conveniently forgetting that there are *many* aspects of change that are open to measurement, not only their duration in time, and so the description of time as 'the' measure of change is quite inadequate by itself. It is also worth noting that time can surely be recognized as a *dimension* of change, distinct from several other aspects of change, even without supposing that the temporal aspect is *measurable* in any meaningful way.[24]

To conclude, I think that we are still missing a good explanation of why Aristotle should think that time is properly defined as 'a number'. I have said nothing in this Additional Note of why he might think that (the? or a?) now can also be regarded as a number. One would hope that a good answer to the first question would bring in its train a natural answer to the second. The essay reprinted here does *propose* answers. I confess that I now find it difficult to believe them, but at the same time I do not know how to improve upon them.

Additional Note 9.2. Before and After

In her (1984) Waterlow supposed that Aristotle means to *define* the 'before' and 'after' in time when he says in chapter 14 that in the case of time past what is before is further from now than what is after, whereas in the case of time to come

[23] Time is described as the measure of change quite frequently in chapter 12, explicitly at 220^b32, 221^b7 and 25, and often elsewhere by implication. But the same chapter also calls time a number (e.g. at 221^a13), and it apparently regards these claims as interchangeable. But elsewhere time is always characterized as a number rather than a measure (save possibly by implication at 223^b12–23).

[24] Time (as commonly conceived) has a definite order and topology irrespective of whether there is any suitable metric for it. Essentially the same point is put in rather different words by White (1992: 86 ff.).

the reverse applies, and what is before is nearer to now than what is after (223^{a}4–13). Since 'before' and 'after' play a central role in Aristotle's account right from the beginning, it seems to me improbable that he would have postponed his *definition* of them to the series of appendices which forms chapter 14. Moreover, if this had been intended as a definition it would clearly have been only a partial one, (*a*) because it fails to say that what is now past is before what is now future, and (*b*) because if we take it *au pied de la lettre* it will follow that what is happening now is neither before nor after any other event, which is surely not what Aristotle intended. (But much of Waterlow's 1984 depends upon supposing that it *is* intended.)

I take it that the crucial definition of time that we find in chapter 11 presumes that we already understand this notion of 'before' and 'after'. The precise sense that he attaches to these notions is clarified by the elucidation which makes them point-like rather than stretch-like, but Aristotle is taking it for granted that we do have a basic understanding of these notions in advance of any explicit definition. It is true that he does offer a further elucidation, which one has to dismiss as just a mistake.[25] He claims that we *first* understand 'before' and 'after' with respect to *place*, and the understanding of their application to movement, and to time, derives from that (219^{a}14–19). But in fact the notions have no independent application to place. No doubt it is fair to say that we can grasp such spatial relations as 'to the left of' and 'to the right of' independently of any understanding of time, and similarly such relations as 'nearer to here' and 'further from here'. But which of these is to count as 'before' and which as 'after' is so far quite undetermined. For this one needs the notion of a *path* in space and of a *direction* along that path. But the most natural way to understand a direction—e.g. from left to right rather than right to left, or spreading out from here rather than closing in on here—is surely in terms of a movement which follows that direction. So it is much more natural to say that 'before' and 'after' are first understood in terms of movement, for that is where the notion of a direction comes from. In space itself there are no preferred directions, but time does have a direction (as movement does), and you cannot understand 'before' and 'after' without grasping that.

25 I regret that reasons of space prevented the proper development of this point in the essay as printed. (It is mentioned only very briefly in the text to n. 16.) But many others have drawn attention to it.

10

Aristotle on Continuity in *Physics* VI

Aristotle's favourite model of the continuum is the same as ours, namely a geometrical line, or line-segment. He does not also have, as we do, the numerical model of the real numbers, but that is scarcely a handicap. After all, one of our main ways of understanding the structure of the real numbers (namely via the notion of a 'Dedekind cut') is very naturally viewed as drawing upon our prior grasp of the structure of a geometrical line, with the real numbers understood as corresponding to the points on that line. There are some features of this structure which Aristotle grasps very clearly, notably that no two points on a line are ever *next* to one another, for between any two points there is always a whole line-segment, which in turn will always contain further points within it. Undoubtedly, this is a major achievement on his part, but its importance should not be over-exaggerated, as it appears to be (for example) in Ross's classic edition of the *Physics*.[1] (Looking at the issue from our contemporary point of view, we may observe that the fact that between any two points there are always others does not yet distinguish the structure of the real numbers from that of the rational numbers.) Ross's general evaluation of the discussion in the first two chapters of *Physics* VI is that 'it seems to me to indicate that Aristotle had a more mathematical turn of mind than he is usually credited with' (p. 70), and his commentary on these chapters contains no important criticism.[2] As for the remainder of book VI, this he says 'develops with unwearied diligence and (so far as I can judge) with unfailing accuracy the implications of the infinite divisibility of time and of extension' (p. 71). We shall see, however, that Aristotle's discussion in fact leaves a number of important gaps, is at times seriously muddled, and contains some plain mistakes. I aim to make these inadequacies clear, and then to add some brief speculations on why they are present.

There is one significant feature of Aristotle's whole discussion in book VI, which I would not regard as an 'inadequacy', but which deserves a mention here

Reprinted from L. Judson (ed), *Aristotle's Physics: A Collection of Essays* (Oxford: Clarendron Press, 1991), 179–22.

[1] Ross (1936).

[2] Ross does note, on 233^{a}18–21, that it is a little surprising to find Aristotle asserting that if time is infinite in extent then so is distance, since his own position is that time is infinite but space is not. The difficulty is removed by recognizing that throughout this chapter Aristotle is speaking of the time taken, and distance covered, by a body in uniform rectilinear motion.

at the outset. That is the tension between Aristotle's reasoning in this book and his discussion of infinity in book III. To put it very roughly, his position in book III is that there is no 'actual' or 'completed' infinite, and all infinity is merely 'potential'. But in book VI he shows no such tendency to be suspicious of the infinite, and apparently accepts without qualms that a line does ('actually') contain infinitely many points, that a stretch of time does ('actually') contain infinitely many instants, and so on. Thus his answer in this book to Zeno's most famous paradox on motion apparently accepts that one who traverses a finite distance has thereby completed a series of infinitely many distinct tasks, in traversing the infinitely many 'half-distances' contained within the original distance. This is possible, he here says, because the finite time available may equally be divided into a corresponding and infinite series of 'half-times' (233^a13–31, 239^b9–29). Later in book VIII he will reconsider this paradox, and with the doctrine of book III in mind he will deny that a moving body ever does pass over infinitely many 'actual' points (263^a4–$^b9$). But here in book VI we find no such complications. A simple hypothesis evidently suggests itself: book VI was written before the doctrine of book III was worked out.[3] (In my concluding section I shall note some other pointers to the early date of book VI.)

1. THE DEFINITION OF A CONTINUUM

In chapter 3 of book V Aristotle defines what it is for one thing to be continuous with another. He begins by defining 'succession' (τὸ ἐφεξῆς): one thing succeeds another when it comes after it, in position or in some other way, and there is nothing between them that is of the same kind (226^b34–227^a6). A special case of succession is 'being next to' (τό ἐχόμενον): one thing is next to another when it succeeds it and is also in contact with it (227^a6), contact being defined as occurring when (some part of) the 'limits'[4] of the two things are in exactly the same place[5] (226^b21–3). A special case of being next to is 'continuity'

[3] Books VI and III do share similar mistakes about infinity. In VI. 7 Aristotle argues that even an accelerating or decelerating object cannot be moving in such a way that it covers a finite distance in an infinite time (237^b34–238^a19), or an infinite distance in a finite time (238^a20–31). Basically his mistake is that he assumes that an infinite time, or distance, cannot be divided into one finite and one infinite part (238^a9–11). This may be compared with his claim in III. 5, that an infinite body cannot have several infinite parts (204^a25–6; cf. 204^b19–22). It may be that one of these errors is influencing the other, though there is no strong reason to suppose so. (Note that III. 5 raises no *logical* objection to the division of an infinite whole into one finite and *one* infinite part, 204^b13–19.)

[4] ἄκρα, 226^b23 and 227^a22–5; πέρατα, 227^a12; ἔσχατα, 227^a13.

[5] ἅμα as defined at 226^b21–2. I take it that Aristotle means what he says, and is (rightly?) supposing that when the face of one (perfect) cube touches the face of another the two faces are different things but are in *exactly* the same place. (By contrast, *Metaph. B.* 5, 1002^a34–$^b3$, appears to suppose that when two such faces touch they automatically do become one, and are no longer two.) Admittedly, this perfectly proper use of 'same place' is somewhat difficult to reconcile with Aristotle's definition of 'place' in book IV, which seems not to allow such things as limits to have places.

(τὸ συνεχές): one thing is continuous with another when they are next to one another and in addition the limits at which they touch are the same thing, or 'have become one', being indeed 'held together' as the word συν-εχές implies. Aristotle's thought evidently is that things which are merely next to one another need not hold together—one may move one of them leaving the other where it is—whereas if they are continuous then they move together (227^a10–17).

At the opening of book VI Aristotle recalls this definition,[6] and appears to suppose that it has also explained to us what it is for a single thing to be a continuous thing—or, as I shall say, a continuum—though clearly it has not explained this at all (231^a21–6). But if we follow through the argument that he at once plunges into, we can I think reconstruct the definition that he has failed to state but must have had in mind. For he claims that it follows from the definition that a continuum (such as a line) cannot be made up from indivisible things (such as points). The first argument is simply that an indivisible thing has no limits, for a limit (ἔσχατον) must be different from what it limits, and thus what has limits must have parts, and must therefore be divisible (231^a25–9). As a second argument Aristotle adds that indivisible things such as points cannot form a continuum by being in contact with one another, even if we waive the point about their lack of limits. For, he says, if a point could be in contact with a point then they would have to be in contact 'as wholes', and this (I think) he takes to imply that they would have to be in exactly the same place. At any rate he objects that contact of this kind will not produce anything continuous[7] because a continuous thing will be divisible into parts that occupy *different* places[8] (231^a29–$^b6$). (The thought here seems to be this. When one (perfect) sphere touches another there will be a point on the surface of the one that touches a point on the surface of the other, and these are different points, just as the face of one cube in contact with the face of another are different faces (see n. 5 above). Hence one point may be said to be in contact with another, but such contact yields spatial coincidence, so no matter how many points we may put together that are in contact with one another in this way we shall still not form anything bigger than a single point. But this subtlety is generally ignored in what follows, where it is simply claimed that indivisible things cannot be in contact (e.g. 231^b17).) Finally he adds as a third argument that points cannot even be successive, since between any two points there will be a line (and hence, he presumably means to add, a further point) (231^b6–10). Summing up this

[6] As Ross notes (1936: 626–7) being next to is defined in book V as a *special* case of contact, namely where the items in contact are also successive. (Compare Plato, *Parmenides* 148e4–7, 149a4–6.) But usually Aristotle draws no distinction between contact and being next to, and simply equates the two, as he does here at the beginning of book VI.

[7] ὅλον δ'ὅλου ἁπτόμενον οὐκ ἔσται συνεχές (231^b4). I take ἔσται to be existential, and ὅλον . . . ἁπτόμενον to be what Ross calls a 'loose accusative absolute'.

[8] τόπῳ κεχωρισμένα. Earlier in book V χωρίς was defined simply as meaning 'in a different place' (226^b22–3). I imagine that what Aristotle has in mind here is places that do not overlap but do touch.

reasoning,[9] Aristotle concludes that what a thing is made up from it may also be divided into, and so we have shown that a continuum cannot be divided into indivisibles. Anything whatever must be either divisible or indivisible, and if divisible then either divisible into indivisibles or divisible only into what may always be divided further; this last is a continuum. And conversely every continuum is divisible into what may always be divided further, for if it could be divided into indivisibles then one indivisible could touch another, 'for the limits of (two) continuous things touch, and are one' (231^{b}10–18).

Now, if this argument is to be intelligible, what must we take the definition of a continuum to be? First, it must evidently be given as a premiss that a continuum has at least two parts which do not coincide (231^{b}4–6). For Aristotle clearly does not count a point as *itself* a continuum, and this seems to be the minimal premiss needed to rule out that suggestion. To avoid some complications I should like to strengthen this premiss a little to the following: a continuum may always be *divided* (without remainder) into two parts that do not coincide. It is true that, given a 'logic of parts' that is nowadays orthodox, the supposedly stronger version would follow at once from the original. For the now orthodox logic supposes that, given any one proper part of a thing, there will always be some *one* further part of the thing, which comprises all the rest of it excluding the given part. Thus the whole may be said to be divided into these two parts in the sense that the two do not overlap one another—i.e. they have no common part—and together they exhaust the original whole, in so far as every part of it must overlap at least one of these two. But it seems to me doubtful that Aristotle would accept the assumption on which this reasoning is based. For example, it implies that if we begin with a line 3 inches long, and subtract a part 1 inch long from the middle of it, then the two separated line-segments remaining may be counted as together forming *one* part of the original. But this is not a very natural use of the notion of one part, and equally it is not very natural to say that the operation just envisaged would divide our 3-inch line into just two parts. I shall come back to this problem shortly, but for the moment let us simply sidestep it, by adopting the stronger premiss: any continuum may always be divided into two.

Then it is perfectly clear from the whole run of Aristotle's discussion that we must add a further premiss: any division of a continuum into two (non-coincident) parts must divide it into parts that are continuous with one another in the sense defined in book V, i.e. the two parts must touch, and the limits where they touch must 'be one'. These two premisses are, I imagine, the premisses that Aristotle regards his argument in this passage as depending upon, and as given by the (unstated) definition of what a continuum is. We may therefore

[9] Ross sees the reasoning from ^{b}10 to ^{b}15 as still concerned with the suggestion that points might be successive. This is mainly because, though he brackets the words τῶν στιγμῶν καὶ τῶν νῦν οὐθέν at ^{b}13, he takes them to be a *correct* gloss. But my summary presumes that they give quite the wrong sense.

reconstruct the definition by designing it to yield just these premisses: a continuum is anything which (i) can be divided into two parts, and (ii) is such that any two parts into which it is divided must share a limit. But if this is indeed the definition Aristotle has in mind, then either it is incorrect or the conclusion that he attempts to deduce from it will not follow. That is, it will not follow that the parts into which a continuum may be divided must themselves be further divisible.

Suppose we take a finite line, and suggest 'dividing' it into these two parts: one part is to be an end-point of the line, and the other part is to be all the rest of the line, excluding this end-point. (Thus the second part is what we call a 'half-open' interval, containing all the points of the line except this one end-point.[10]) Now we may surely assume that a finite line is a continuum, if anything is, so if Aristotle's definition is correct then the two parts into which we have divided it must share a limit. According to one of Aristotle's lines of thought, they do not, since he claims that a point has no limit. But in that case we must simply reject the definition as incorrect. However, if my interpretation is right, then he also has another line of thought which allows points to be in contact with one another ('as wholes') despite their alleged lack of limits. This is most easily harmonized with his general position by allowing that a point may be said to be *its own* limit. If we do allow this, then the suggested definition of a continuum may be retained, even in the face of this example. For the end-point of the line, and the rest of the line, now do share a limit, namely the end-point itself. It is its own limit, and it must also be reckoned to be one of the limits of the rest of the line. For we cannot say that the rest of the line has no limit in this direction without once more subverting the suggested definition, and if it is to have a limit then clearly there is no other candidate than this same end-point. Thus, if the definition is to be retained, while admitting this example as a 'division', then we must conclude that it is possible to divide a line into two parts, where *one* of these parts is a mere point, and hence not further divisible. (And we may add, incidentally, that *any* point of the line may thus be exhibited as one of the parts into which it may be divided, by considering a division into three parts, one of which is the given point, one the rest of the line (if any) that is to the left of it, and one the rest of the line (if any) that is to the right of it.)

There are two possible avenues one might explore, in seeking to rescue Aristotle from this objection. One would be to note that I have merely conjectured his definition of a continuum (since he fails to state any definition himself), and to seek for an alternative conjecture. Here one may remark that he quite often appears to take, as *definitive* of a continuum, the thesis that in this passage he is attempting to prove. That is, a continuum is defined as what is

[10] One might claim that Aristotle himself may be said to recognize the existence of such things as half-open intervals at *Phys.* VIII. 8, 263^{b}9–264^{a}6, where he perceptively remarks that an interval without its end-point is no shorter than the interval with its end-point (264^{a}4–6).

divisible (only?) into parts that are further divisible. (This appears to be cited as a definition a little later in this very book of the *Physics*, at 232^{b}24–5; it is also cited as a definition elsewhere, e.g. *De Caelo* 268^{a}6–7.) But there are two obvious objections to any such suggestion. One, based on the passage we have just been considering, is that such a definition simply omits a feature that Aristotle evidently regards as crucial: it does not include the condition that the parts into which a continuum may be divided must share limits. The other, based on the objection that we have just been considering, is that any definition designed to rule out that objection would run the risk of having as a consequence that a geometrical line is not after all a continuum. But this consequence is evidently intolerable. The alternative escape route is to put some restrictions on what is to count as a 'division' so that the example we have considered does not qualify. It is not easy to see quite what restriction would be appropriate, but whatever is suggested there will always be this reply: if we are only allowed to count as a division something which divides a continuum into extended parts, each further divisible, then it will no longer be plausible to claim (as at 231^{b}10–11) that whatever a continuum is 'made up of' it will also be 'divisible into'. That is, a line may perhaps be constructed out of points, and nothing else, even though it is not 'divisible' into points, according to the restricted understanding of a 'division'. I conclude that neither of these avenues of escape will in the end prove satisfactory.

But here we should notice a way of modifying Aristotle's argument, which still yields the conclusion he was aiming for, and derives this conclusion perfectly correctly. It requires, however, a much stronger premiss on divisibility. We have said so far that any way of dividing the line into just two parts must yield parts that share a limit (and we have observed that one of those parts may be taken to be a mere point without disturbing this premiss). By repeating such divisions, we can evidently infer that any way of dividing the line into a finite number of parts must yield parts that share limits (and several of those may be taken to be mere points, so long as others are line-segments joining them). But the generalization that Aristotle's argument requires is that any way of dividing the line into any number of parts—finite or infinite—must yield parts that share limits. However, no point shares a limit with any other point, and hence the line cannot be divided into *nothing but* points. But what a thing is made up of it may also be divided into, and hence a line cannot be made up of nothing but points. This, evidently, is an argument of some power, and it is not at all surprising that Aristotle found it compelling, but of course we are free to reject the strong premiss on divisibility that it is based on.

But still, though we may reject Aristotle's premiss, we can yet feel some sympathy with his position. How *can* we conceive of a line as made up of nothing but points, which cannot touch one another? Imagine that you have an inexhaustible supply of points to hand; how could you 'put them together' (as it were, 'one at a time') so as to make a line out of them? Well, we can say nowadays that the answer is this. First set down a denumerable infinity of points that is dense in the line, for example by putting down the two end-points of the desired line, and then by setting down a further point in the middle of any stretch between two points already set down. (Note that we do *not* try to put the points down 'in succession' from left to right.) When you have finished this (!), then consider all the Dedekind cuts in the points already set down, in other words all ways of separating those points into two groups, with all of one group to the left of all of the other. For each such cut, put in a further point to be the point at which the separation is made (if you do not have one already—i.e. if there is not either a rightmost point of the left group, or a leftmost point of the right group). We may note, incidentally, that to complete this stage you will in fact need more than a denumerable infinity of points (so it may take you a little longer than the first stage), but when you have done it you are through. The result is now a continuous line. But although we have a proof of this fact, must we not still admit that it is absolutely amazing? No point that has been put down touches any of the others, and yet the result is a line with no gaps in it anywhere! No wonder that Aristotle could not see how such a construction of the line could succeed. Indeed, I think I would admit that I cannot exactly 'see' it myself.

So much, then, for Aristotle's opening argument in book VI, as seen from a modern perspective. I now add two observations (from the same perspective) about the definition on which it is based. The first follows naturally from the mention of Dedekind cuts in the last paragraph. Elsewhere in *Physics* VI Aristotle is generally content to characterize a continuum as 'divisible into divisibles' (and hence 'infinitely divisible'), but he pays surprisingly little attention—either here or elsewhere—to the question of how much divisibility a continuum has. For example, when he is arguing that time is a continuum, he seems content with a proof that any time, however short, can be divided to yield parts that are yet shorter (232^{b}20–233^{a}12). He is prepared to add that any time can be divided in a specified ratio (233^{a}3–5), or again that any time can be divided both into two equal parts and into three equal parts (233^{b}19–32), but these points seem merely incidental to his main claim, which is just that time is infinitely divisible. One might have expected him to stress that a continuum can always be divided in *any* desired ratio, not only those ratios that can be specified by the natural numbers, but also those irrational ratios that geometry forces upon us, e.g. the ratio of $\sqrt{2}$ to 1. (These, of course, correspond to the Dedekind cuts mentioned above.) But in fact there is no emphasis on this point, and indeed no mention of irrational ratios in the whole of book VI. I can only comment that I find this surprising. As a matter of fact it can be proved, from what I at least regard as plausible assumptions, that if there is no smallest time then it does *follow* that any time may be divided in any desired ratio, rational or irrational.[11] But one can hardly suppose that Aristotle knew of any such proof. As for the characteristic of a

[11] See e.g. my (1979: 173–7, 237–9, 248–53). But my 'plausible assumptions' include a principle that provides for the existence of Dedekind cuts (see in particular pp. 270–4).

continuum that Aristotle stresses here, concerning the sharing of limits, this does indeed have a role to play in the modern conception, as elaborated in topology. I end this section with a few brief and sketchy remarks on this.

In modern topology one takes as primitive and undefined the notion of an open set of points. This, one might say, corresponds to the notion of something that is extended, that fills space. We may remark here that ordinary extended things will be extended in one, two, or three dimensions, and the dimensions will introduce an order, allowing us to speak of a series of extended things, coming one after another in the order corresponding to that dimension (e.g. from left to right, from top to bottom, from earlier to later). This explains the relevance of Aristotle's notion of succession, which receives little attention in his own account. But indeed it is not of fundamental importance. What is much more important is that an extended thing will (normally) have a limit, which of course corresponds to the points in the closure of the corresponding open set, but not in the set itself. For most purposes, it is natural to think of an extended thing as including its limits, but we have now learnt how important it is to be clear about this, one way or the other. Now any extended thing may be divided into two others (of the same dimensions), and this corresponds to the partition of an open set into two open subsets, plus the points, if any, that are in the limit of both. But in the general case there need not be any points in the limit of both, for an extended thing may consist of several scattered parts, not in any way joined up with another. That is to say, an open set may be a union of open sets that are not in any way *connected* with one another. This is where Aristotle's definition comes in: it introduces the requirement of connectedness. So a continuum is a special case of an extended thing, namely one where any two parts into which it may be divided must share a limit.

Now the notion of connectedness proves, on investigation, to be very much more complicated than one might at first have supposed. (For example, we need to distinguish between being connected 'pathwise' and being connected in other ways.) That is why Aristotle's definition is not the one used in modern topology, though I cannot elaborate here on the issues involved. But from our perspective it can easily be seen as an attempt to introduce just this notion; its role is to distinguish extended things which are 'all joined up' from those that consist of scattered parts. But then one must observe, with some irony, that it seems improbable that Aristotle himself was aware of this role, since it seems improbable that he recognized the existence of any extended things that were not continua.

2. THE CONTINUITY OF SPACE AND TIME

For the remainder of chapter 1 of book VI (from $231^{b}18$), and for all of chapter 2, Aristotle's aim is to show that time and space are each infinitely divisible, and hence continua.[12] His argument depends upon the premiss that motion is also continuous, which he explains in the same way as meaning that any motion is always divisible into further parts that are motions (231^{b}18–20), but this idea needs a little elaboration.[13] Throughout the discussion in these chapters, Aristotle is considering only uniform rectilinear motions, covering equal distances in equal times, and he counts such a motion as continous if the moving body traces out a path through space that is a continuous line. That is, each point on the line is occupied by the body at some instant during the time of its movement. The alternative that he considers is that motion proceeds, as we might say, 'cinematographically' (or, as Aristotle says, *ἐκ κινημάτων* rather than *ἐκ κινήσεων*, 232^{a}8–9), which is to be envisaged thus: the moving body remains at one position for a stretch of time, and then 'leaps' instantaneously to another and different position, without ever occupying the positions in between. It then rests at the new position for a further stretch of time, then 'leaps' instantaneously to yet another position, and so on. Aristotle argues against this alternative hypothesis at 231^{b}28–232^{a}18.

Suppose we have a spatial stretch *A* and our allegedly moving body 'moves' across it by staying for a while at one end of it and then instantaneously leaping to the other. Then Aristotle in effect claims that there is no time when the body *is moving* across *A*; there is a time when it is not yet moving across *A*, and there is a time when it has moved across, but no time when it is crossing. But, he says, one cannot be in a position in which one *has* walked to Thebes without at some time being in the process of walking to it. And it is no use, he suggests, to suppose that a longish journey, such as that from Athens to Thebes, is actually made up of a whole series of such small instantaneous leaps, so that I can say that I *was* walking to Thebes during all the time when that whole series was going on. For that supposed movement would be made up of things that clearly were not movements, but states of rest (before an instantaneous transition) and states of *having* moved (after it). But a movement, he claims, can only be made up of movements, just as a line can only be made up of lines and not of points; no other alternative is possible.

To this one can only reply that the alternative that Aristotle himself quite clearly describes in this passage is evidently not impossible at all. No doubt we do not believe that movement takes place through such a series of instantaneous leaps, separated by rests, and for that reason our usual vocabulary for talking of movement is not well adapted to this possibility. But that is surely of no significance. It is simply an empirical question whether movement is in this way discontinuous, and one that is hardly settled by the observation that it does not look that way to the naked eye. But if the continuity of movement is best

[12] Ross's title for chapter 2, viz. 'Further Proof that Every Continuum is Infinitely Divisible', is thus misleading.

[13] From our point of view, the continuity of motion is a matter of the continuity of a *function*, namely the function which assigns to each instant during the motion the position occupied at that instant.

regarded as a straightforwardly empirical question, should we say the same of the continuity of space and of time?

What is undoubtedly true is that all of Aristotle's arguments for the infinite divisibility of space and time do depend upon the premiss that motion is continuous, and will collapse if that premiss is denied. I illustrate this claim by considering just his final argument at 233^b15–31, for (unlike the others) that argument appears at first sight to be independent of the continuity of motion. We may reasonably suppose that two bodies may move (uniformly) in such a way that the ratio between their velocities is as three to two. We may also suppose that if there are indivisible time-atoms and space-atoms[14] then the faster body is moving at such a rate that it traverses exactly one space-atom in exactly one time-atom. Thus after three time-atoms it has crossed three space-atoms, and in the same time the slower body will have crossed two space-atoms. Then, Aristotle argues, the slower body will cross one space-atom in half of three time-atoms, which is only possible if the supposed time-atoms are after all divisible, and hence not atoms at all.

Now, if we assume that the motions in question are continuous, this is no doubt a fair conclusion. But if we suppose instead that the motions are cinematographic then the conclusion is easily avoided. The faster body, we may assume, proceeds by remaining at rest for one time-atom and then instantaneously leaping a distance of one space-atom. The slower body must proceed in a manner which has the result that it has crossed two space-atoms in three time-atoms. Evidently the simplest hypothesis is this: it proceeds by remaining at rest for three whole time-atoms and then instantaneously leaping a distance of two whole space-atoms. We do not have to suppose that in cinematographic motion a 'moving' body never rests for more than one time-atom, nor that it never leaps more than one space-atom. We can make the first supposition if we wish to, and preserve the atomicity of time, while allowing the infinite divisibility of space. Equally, we can make the second supposition, and preserve the atomicity of space, while admitting the infinite divisibility of time. But we do not have to make either supposition. So we can if we wish maintain the atomicity of both space and time, so long as we also claim that the ratio of one 'uniform' velocity to another can always be measured by a pair of whole numbers. And clearly no straightforward empirical observation could provide a direct refutation of that claim.

It will be useful to add a word here on how one should conceive this hypothesis of the 'atomicity' of space and time. The basic idea is that although time, say, *may* be viewed as forming a continuum, such a view turns out to contain superfluous detail, in that it posits the existence of many temporal instants which

[14] I shall try to elucidate this notion in a moment. Meanwhile, observe that in my usage an atom is extended, and may even be said to have parts, though there is a sense in which it is not divisible.

need never be invoked in the explanation of physical phenomena. For the only instants at which any change ever occurs are the instants that separate each time-atom from the next. Thus the time-atoms are conceived as having duration, and may be said to be 'divisible in thought', so that we may speak of the state of affairs that obtains 'halfway through' a time-atom. (Indeed, the claim will be that exactly the same state of affairs persists all through the time-atom.) But this 'divisibility in thought' corresponds to no reality, and is merely a product of our way of conceiving of time, namely as a continuum measurable by the real numbers. Perhaps, if we had reason to suppose that time really was atomic in the way suggested, we might (for scientific purposes) adopt a different way of looking at it, taking it as a matter of definition that each time-atom was equal to every other, and thus measuring time by the natural numbers (used simply to *count* the time-atoms), rather than by the real numbers. (One might compare the way that we adopt, for scientific purposes, an 'absolute' scale of temperature, which eliminates those degrees of temperature, below − 273 °C, to which no reality corresponds.)

Certainly, we could not adopt such a standpoint without abandoning many deeply ingrained ways of thinking. Quite a nice illustration of this is to be found in Aristotle's argument in chapter 3. He claims there that what he calls 'a now, in the primary sense' must be conceived as an indivisible instant and not a stretch of time, offering this argument. There will be an instant which is the limit of time past, and an instant which is the limit of time to come, and we cannot suppose that 'the present time', i.e. 'now', is some stretch of time between these instants. If we try to think of it in this way, then we must admit that this stretch is divisible, and then a (suitably chosen) division would itself have to be 'the now, in the primary sense', dividing the past from the future (233^b33–234^a24). Putting his argument the other way about, we may say that his point is that if we try to conceive of an 'atomic present', which lasts for some time, but which is 'atomic' in the sense that nothing changes during that time, then we will find an insuperable obstacle. For at least *the time itself* must be conceived as changing. That is, the limit of time past must be envisaged as moving continuously through the supposed atomic present, and cannot be thought to rest at one end of it and then leap instantaneously to the other. But the reply is simply that although it may indeed be difficult to conceive the 'flow' of time in this way, it would not appear to be impossible. (And we may add, *ad hominem*, that Aristotle's own doctrine that time depends upon change would require us to say that time itself cannot be continuous unless some *other* change is also continuous.)

To generalize, scientific theories of this century—most notably the theory of relativity—have taught us that many common-sense assumptions about the nature of space and time are in error, and should be abandoned. The continuity of space and time might perhaps be another such common-sense assumption, which we will eventually have to abandon. One can at least roughly envisage developments of quantum theory which would point in this direction, and the rival theory of 'atomicity' is perfectly coherent. But of course we all nowadays accept Aristotle's view that space and time are continua, and that motion too is in its way continuous, and I too shall accept these views in what follows.

3. STATES AT AN INSTANT

Much of Aristotle's discussion for the remainder of book VI is concerned with what may be true of a thing at an instant. He begins on this during the rest of chapter 3 (from 234ᵃ24), with an argument that purports to show that there can be no motion or rest at an instant.

The claim as he first phrases it is that 'nothing moves in an instant' (*οὐθὲν ἐν τῷ νῦν κινεῖται*), and his first argument for it is evidently that nothing can cover any stretch of distance during an instant (234ᵃ22–31). This is so far quite compatible with our being able to say that a thing *is moving* at this or that instant, but it soon becomes clear that Aristotle wishes to rule this out too. We find that something odd is going on in the next sentence, where he claims that just as nothing moves in an instant so also nothing rests in an instant, his ground being that one can only speak of rest where motion is equally possible (234ᵃ31–4). But clearly the argument that he has just given for denying motion in an instant would not directly apply to rest. However, we have two more arguments to come. In the first, he asks us to consider a stretch of time divided by an instant into two sub-stretches, with a body that is moving throughout the whole of one sub-stretch and at rest throughout the whole of the other. Then, he says, since the dividing instant is 'in' both sub-stretches, if we could speak of the body moving or resting at that instant we should have to say that it was both moving and at rest then. This he evidently regards as an impossible result (234ᵃ34–ᵇ5). Finally, he argues that to be at rest is to be in the same state 'both now and earlier', and there is no 'earlier' in a single instant, and hence no rest then either (234ᵇ5–7).

It is clear that these arguments are wholly inadequate. In reply to the first one might say that there may perhaps be a special problem over the instant of change from rest to motion or from motion to rest, but there is no such problem about any other instant in our stretch. We may surely say that the body is moving at any instant that falls within the one sub-stretch, and is at rest at any instant that falls within the other. Moreover, the final consideration appears to give us a way of resolving the problem over the instant of change itself, for it evidently directs us to say that the instant of change from rest to motion is an instant of rest. If we may add that there is a further way of being at rest, namely to be in the same state 'both now and later', then we may similarly infer that the instant of change from motion to rest is also an instant of rest. Thus there will be first and last instants of rest, but no first or last instants of motion. All that is needed is a *decision* as to what to say about these special instants of change; and it is surely an extravagant reaction to ban *all* talk of a thing's state of motion or rest at an instant. (I add, incidentally, that the decision just suggested appears to be a perfectly reasonable one, and may be further supported by a point that I take from Sorabji.[15] If we may assume that all changes of velocity are continuous, then clearly the instantaneous velocity of an object at the instant of change from rest to motion, or motion to rest, must be zero.)

We get no further argument on this issue in *Physics* VI (nor, I think, elsewhere in Aristotle). The arguments we have considered are repeated in chapter 8 at 239ᵃ10–22, but no new point emerges. One thing that we do find is that Aristotle intends his conclusion to apply, not just to motion proper, but to change of all kinds: one is not permitted to say, of any instant, that a thing is changing at that instant (237ᵃ14). It should also be noted that at one point in chapter 5 Aristotle himself seems to speak of a (last) instant of rest (236ᵃ18), but this must presumably be taken to be either a slip on his part, or (more probably) a claim that he thinks that his opponent in that passage—namely one who believes in time-atoms—would have to grant.[16] It is clear that his considered doctrine is that there is no rest, movement, or other change, at an instant. But what is difficult is to reconcile this view with some further claims that he goes on to make about what *can* be said to hold at an instant.

In chapter 5 he argues in effect that if a thing changes from being in state *A* to being in state *B*, then there is a first instant at which it *has* changed into state *B*, i.e. has completed its change. Evidently, this instant must be the same instant as that which separates its not being in state *B* from its being in state *B*, and so Aristotle's claim is that at that instant of change it counts as having reached state *B*, and therefore is then *in* state *B* (235ᵇ6–236ᵃ7). In contrast to this, he goes on to argue that there is no first instant at which the thing has begun its change, or has changed at least a bit (236ᵃ7–27), and this theme is elaborated further at the beginning of the next chapter (236ᵇ19–32). It is evident from this that the changes Aristotle is considering here are changes that take place over a whole stretch of time. Let us simply accept this for the time being. (I shall consider it more closely in my next section.)

Now the obvious difficulty with this is that a change from, say, travelling at 10 miles an hour to being at rest, or vice versa, would seem to be included in the discussion. But then we must say, according to the doctrine propounded, that there is a first instant at which this change is completed, i.e. a first instant at which the thing has come to rest, or has reached the speed of 10 miles an hour. And it seems very natural to add that an instant at which a thing *has* come to rest, or *has* reached a certain speed, will also be an instant at which the thing *is* at rest, or *is* moving with that speed. But of course if we do say this then we are in flat contradiction with what has been argued earlier.

[15] Sorabji (1976: 69–89).

[16] The second suggestion is due to Sorabji (1983: 415 n. 17). This is a recantation of the view put forward in his earlier article (1976). I discuss the suggestion in Section 6 below.

It is not easy to see how Aristotle proposes to avoid this contradiction. It is not that he wishes to deny the inference from 'has come to be in state *B*' to 'is in state *B*'. Indeed chapter 5 opens with an argument *for* this inference (235ᵇ6–30), and equally when he recapitulates the doctrine in chapter 8 of book VIII he very clearly accepts that at the first instant at which a thing has come to be white, at that instant it is white (263ᵇ9–264ᵃ6). Moreover, he explicitly accepts the inference for the particular case of coming to be at rest, a topic to which he devotes chapter 8 of book VI. Coming to rest, he first argues, is a process which takes time, though there is no smallest time in which it takes place (238ᵇ23–239ᵃ10). But then he adds that there is no first instant at which a thing *has* come to rest, precisely because that would imply an instant at which it *is* at rest, and this he has denied (239ᵃ10–14). (And he goes on to remind us of his reasons, as stated earlier in chapter 3, 239ᵃ14–17.) So he then infers that being at rest is also something that takes time, and again that there is no first time in which it occurs. If we put this together with the doctrine of chapter 5, that for *every* kind of change there is a first instant at which the thing has reached the terminal state of the change, then we can only infer that in Aristotle's view a change from moving at a given speed to being at rest (or vice versa) does not count *as* a 'change' (μεταβολή). The apparently restrictive doctrine of chapter 2 of book V, which argues that all change can be classified under just four headings, may indeed have been intended to point forward to this result. But we may still complain that it is without adequate motivation.

I add that it is of course perfectly common usage to speak of a thing as moving with a given speed, or as being at rest, *at* this or that instant, and it is a usage which Aristotle himself falls into without noticing it. His thesis that whatever is changing has changed, i.e. has changed at least a bit, is quite naturally expressed by stressing that it has *already* changed, at that very same time at which it is changing. Moreover, Aristotle himself later proceeds to elaborate his thesis further, claiming that what is changing has *previously* changed, i.e. at a time *before* that at which it is changing (236ᵃ34, 237ᵇ5, 10–11, 18, 20). But we cannot take this to mean that it has changed before the *stretch* of time during which it is changing, without evidently falsifying the thesis. What is intended is clearly that for any *instant* at which it is changing there is an earlier instant at which it has changed (or, in the first version, that any instant at which it is changing is also an instant at which it has changed). One has to go a long way round about even to express the thesis that Aristotle intends if one cannot speak of instants at which a thing is changing.

It was, then, fairly clearly a mistake on his part to deny that we can speak of rest, motion, or change at an instant, and not only because the notion of instantaneous velocity is of course crucially important to Newtonian physics. (No doubt, Aristotle would not have found it easy to give an adequate definition of instantaneous velocity, but that is hardly a good reason for denying its existence.) On the contrary, it is a perfectly common-sense notion, and one which it is difficult to avoid when stating Aristotle's own position. Moreover, its denial introduces a quite needless conflict with his thesis that for any change there is always a first instant at which it is completed.

As for this latter thesis, I remark for the moment only that we have been given no reason to accept it. (All that Aristotle actually argues in 235^{b}30–236^{a}7 is that *if* there is a first time in which a thing has changed, then that time must be an instant.) It may, however, seem to be a perfectly reasonable thesis, when what we are considering is a change from being in state *A* to being in state *B*, and this change is one that takes time. For the time taken by the change will then be bounded by two instants, and it will be natural to assume that the second of these will also be the first instant of being in state *B*, and the first of them will also be the last instant of being in state *A*. (Aristotle does explicitly affirm one of these assumptions, as we have noted; as we shall see shortly, he is also committed to the other.) Hence there will be, as also seems plausible, neither a first nor a last instant at which the thing is changing from being in state *A* to being in state *B*. But all this is on the hypothesis that the change itself does last over a stretch of time. Aristotle proceeds to affirm that *all* change is of this kind.

4. ALL CHANGE TAKES TIME

During the first half of chapter 6 Aristotle further elaborates his claim from chapter 5, that there is no first instant at which a thing has begun its change, or has changed at least a bit. He expresses this, as we have noted, as the thesis that if a thing is changing then it has changed (236^{b}19–237^{a}17), and he then proceeds to argue also for the converse, that if a thing has changed then there must have been a previous time during which it was changing (237^{a}17–^{b}9). Now we have already met this claim in chapter 1, as a claim specifically about motion: if a thing has walked to Thebes, then earlier it was walking to Thebes (232^{a}10–11). But while we may perhaps accept that this is true of motion (thus ruling out 'cinematographic' motion), what Aristotle is now urging is that the same holds of every kind of change whatever, and this clearly cannot be accepted.

His argument is this. Suppose that a thing has changed 'in an instant' from being in state *A* to being in state *B*. Then it has not changed to state *B* 'in the same instant in which it is in state *A*'. Yet if it changes (from state *A*?) in any other instant, then there will be a stretch of time between the two, and the change will have been going on throughout that whole stretch (237^{a}20–5). But clearly Aristotle must be assuming here that in a change from state *A* to state *B* there will not only be a first instant of being in state *B* (as has indeed been claimed) but also a last instant of being in state *A*. If this is granted, and assuming (as we may) that nothing can be simultaneously both in state *A* and in state *B*, then indeed Aristotle's conclusion follows: these two instants must be different, and hence there must be a stretch of time separating them. But why should one assume that both these instants must exist? On the contrary, if (as has been claimed) there is a first instant of being in state *B*, then at all instants previous to this the thing is not in state *B*, and we are free to suppose that it is instead in state *A*. But there is no temporal gap between a given instant and all the instants previous to it.[17]

Hoping to clarify things further, Aristotle goes on to say that his claim is clearest when we consider change of position, since the distance between one position and another cannot be indivisible (237ᵃ28–35). And no doubt his claim is in this case perfectly sound (granting the continuity of motion). But he goes on to add that the same applies to *all* changes, even changes between opposites (ἐνάντια) and between contradictories (τὰ ἐν ἀντιφάσει) (237ᵃ35–ᵇ2). However, it is perfectly clear that the same does not apply in these cases. There is no analogue of the 'distance' to be traversed between two contradictory states, and even though there may be an analogue in the case of opposites—e.g. the 'distance' between black and white that is constituted by the various shades of grey, or the 'distance' between low and high that consists of the various intervening tones—still a thing may surely 'leap' from one point on the scale to another without having to run through all the intermediate points. Aristotle's justification for his claim in these cases is that though we may not always be able to divide the 'distance' between the termini of such a change, still we can always divide the time taken (237ᵇ2–3). But this, of course, begs the question at issue. For in these changes between opposites there need not be any time taken, and certainly in a change between contradictories there cannot be. Let us consider the implications of this latter point.

It is a plausible principle that at any instant at which a thing *is changing* from being in state *A* to being in state *B*, it is neither (still) in state *A* nor (yet) in state *B*. But if *A* and *B* are contradictories, then at every instant a thing is either in state *A* or in state *B*, and at no instant is it in both states. It then follows that there is no instant at which the thing is changing. Thus, to use Aristotle's own example, if a thing becomes white, then there are instants at which it is (still) not white, and after all of these there are instants at which it is (already) white, but there is no instant 'in between' at which it is neither white nor not white, and therefore no instant at which it is *becoming white*. Alternatively, if we insist that there must be at least one instant at which it is changing, then this must either be the same instant as the first instant at which it is (already) white—if, as

[17] There is a similar mistake later in chapter 8 of book VIII (262ᵃ28–ᵇ3), where Aristotle argues that if we can say, of a thing in motion, that at some instant it has reached a given point, and later has left it, then we must suppose that there is a *first* instant at which the thing has left the point. From this he infers that since the instant of reaching the point and the supposed instant of leaving it cannot be the same, they must be separated by a temporal gap during which the thing remains at the point. Consequently, if the thing simply passes the point without pausing at it, there is no instant at which it has reached the point. This, incidentally, is in support of his view in book VIII (in conformity with the discussion of infinity in book III) that merely passing a point does not 'actualize' it. By contrast, book VI sees no problem with the idea that there is an instant at which the thing is level with the point, 239ᵃ33–ᵇ4.)

Aristotle affirms, there is such a first instant—or otherwise it must be the same as the last instant at which it is (still) not white, if instead there is such a last instant. So in the first case we must say that at the same instant a thing may both be in state *B* and be changing to it (a thesis that Aristotle explicitly denies, e.g. at 235ᵇ25–6), and in the second that at the same instant a thing may both be in state *A* and be changing from it (a thesis which he also denies, e.g. at 234ᵇ10–15 and at 240ᵇ28–30). So the truth is that change between contradictories simply cannot be made to fit in with the claims that Aristotle here makes.[18]

It is, to say the least, somewhat surprising that Aristotle says nothing at all about this problem. But perhaps one can conjecture his answer to it by looking more closely at what he says in a passage towards the end of chapter 9, where he explicitly claims that change between contradictories presents no difficulty for his account (240ᵃ19–29). He implies, it seems, that it would be a difficulty if what was changing from being not white to being white were in neither state, but he goes on to claim that he is not committed to saying this. For a thing need not be wholly white, or wholly not white, before it can be said to be white or not white respectively: we call a thing white so long as the majority of its parts, or its most important parts, are white (240ᵃ20–5). The same applies, he concludes, to other changes between contradictories: 'for the thing will always be in one or other of the opposing states, but it will not always be in either of them as a whole' (240ᵃ26–9). Now on the face of it this does not resolve our problem, for it reaffirms that at any instant the thing either is white or is not white, and so leaves no time for it to be changing from the one to the other. But at the same time the passage strongly suggests—without quite affirming it—that the change in question is 'really' between contraries and not between contradictories, namely between the initial state of being *wholly* not white, and the final state of being *wholly* white. There will be room for a whole stretch of time for this change to occupy, if—as is clearly envisaged here—the change spreads bit by bit over the surface in question (cf. 236ᵇ6–8, 230ᵇ32–231ᵃ1).

To generalize, then, Aristotle's thought will be that all change can be seen as taking time, so long as we take care to describe the *termini* of the change appropriately. We may loosely describe a change as being from white to not white, or vice versa, and there is admittedly no state intermediate between these. But if we specify the termini precisely—e.g. as wholly white and wholly some other (specified) colour—then there will certainly be room for intermediate states, and hence for a time taken in passing through them. We may also add that care must be taken in specifying the termini of a change if we are to maintain the earlier doctrine that, in a change from state *A* to state *B* which does take time, there will always be a last instant of being in state *A* and a first instant of being in state *B*. For example, suppose I move an object so that its centre of gravity

[18] It strikes me as somewhat ironical that Aristotle cites change between contradictories as the *clearest* support for his contention that there always is a first instant at which a thing has completed its change (235ᵇ16–30).

changes from being to the left of one end of a ruler to being to the right of the other end of it. Then undoubtedly this change takes time. But presumably there is no last instant at which the centre of gravity is to the left of the one point, nor any first instant in which it is to the right of the other. (Rather, there is a first (and only) instant at which the centre of gravity coincides with the one point, and a last (and only) instant at which it coincides with the other.) Hence to maintain the doctrine in a case such as this we must insist that the termini of the change be more precisely specified: it is not good enough to say that at the beginning the centre of gravity is 'to the left of' some given point, but we must say more precisely how far to the left it is (or must specify its precise position in some other definite way). Now it seems to me quite plausible to hold that in any change which does take (a finite[19]) time it will be possible to specify the initial state in such a way that there is a last instant of being in that state, and to specify the final state in such a way that there is a first instant of being in it. So in this revised form the doctrine that there always are both a last instant of the initial state and a first instant of the final state may perhaps be accepted, but only—of course—for changes which genuinely do take time. But we still have not been given any reason to suppose that all changes genuinely do take time, and hence that if a thing *has* changed then there must have been a time during which it *was* changing.

As a matter of fact this claim seems to be abandoned later in the *Physics*, in chapter 3 of book VIII. For it is there claimed that a thing may change size by losing a whole particle of matter all at once—and though the particle is divisible it does not follow that it was detached bit by bit[20]—and similarly any change of quality (such as freezing) may happen all at once (253^{b}14–26; cf. also 186^{a}15–16). On the other hand Aristotle does not there, or anywhere else in the *Physics*, draw the moral that a thing may have changed without ever having been in the process of changing. It is a point that he insists upon elsewhere, frequently in the *Metaphysics*,[21] but it does not seem to appear in the *Physics* itself. It seems *almost* to be recognized in the coda to this same chapter 6 of book VI, where Aristotle turns to apply his doctrine to coming into being and passing away. As one might expect, he claims there that if anything has come into being then at an earlier time it was coming into being, and that at any instant during that earlier time some part of it had already come into being (237^{b}9–22). But this of course requires that the thing coming into being has parts, and indeed is infinitely divisible, and Aristotle does preface his claim with a restriction to such things (237^{b}11). However, he does not here say anything about the coming into being

[19] A change which takes an infinite time presumably will not have both an initial state and a final state, and the doctrine that for every change there is a first instant at which it is completed needs modifying to allow for this. (The eternal rotation of the heavens would be an example; cf. the obscure remarks on this at 265^{a}27–^{b}8)

[20] Contrast 236^{b}16–17.

[21] e.g. *Metaph. B.* 5, 1002^{a}28–^{b}11; *Z.* 15, 1039^{b}20–7; *H.* 5, 1044^{b}21–9; *Λ.* 3, 1070^{a}15–17. Cf. *Cael.* 280^{b}25–8; *EN* 1174^{b}10–13.

of a thing with no parts (e.g. a point), though elsewhere it is one of his favourite examples of what has changed without ever being in the process of change. On the contrary, the reader of book VI of the *Physics* will be more likely to suppose that the restriction is otiose, for Aristotle has already urged, and will urge again, that *everything* that changes in any way must have parts. Let us take up this claim, as our final topic from book VI.

5. WHATEVER CHANGES HAS PARTS

The claim is argued first at the beginning of chapter 4, and the reasoning is this. When a thing changes, it changes from being in one state to being in another, say from being in state *A* to being in state *B*. Then when the thing is in state *B* its change is completed, and it is no longer changing. Equally, when it and all its parts are in state *A*, it is not yet changing. Assuming, then, that there must be a time during which it is changing—and we have just considered the argument for this assumption—Aristotle infers that during that time part of it must be in state *A* and part in state *B*, 'since it cannot be in both of the states, nor in neither of them' (234^b10–17). But it is clear that this is wholly illegitimate reasoning; if I am to change from being in Athens to being in Thebes, it is by no means necessary that during all the time that this change is going on part of me must be still in Athens and part of me must be already in Thebes. Apparently in an attempt to avoid this, and similar, objections, Aristotle proceeds to add that the state *B* is to be taken as 'that into which the thing *first* changes' in the course of its whole change. For example, if the whole change is from being white to being black, then, 'that into which the thing first changes' is being grey (234^b17–20). But here of course we may object that there need not be any such 'first' state, and there cannot be if, as with motion, the states intermediate between *A* and *B* form a continuum. Indeed, Aristotle himself is going to argue for that very point in the next two chapters. So this argument at the beginning of chapter 4 must be dismissed as worthless. (Nevertheless, Aristotle proceeds to elaborate its consequences for the rest of chapter 4—often somewhat obscurely—and ends the chapter by suggesting that it is the divisibility of the changing object that is the foundation for the divisibility of time, space, and movement, 235^b1–4.)

The divisibility of the changing object is assumed several times in what follows (e.g. 236^a27–35, 240^a22–9), and Aristotle comes to argue for it once more at the end of the book in chapter 10. This time he begins with a qualification. A thing without parts, he concedes, may be said to change *per accidens*, in so far as some whole that contains it is changing (240^b8–13). (And the change of the whole, he adds, need not be the same as that of its parts, illustrating this by contrasting the changes of the parts of a rotating sphere, those near the centre moving more slowly than those near the circumference, 240^b13–20.) But a thing without parts, he claims, cannot change *per se*, and at first he appears simply to repeat the argument of chapter 4. For suppose such a thing does change, its initial state being *AB* and its final state *BC*. Then during the time when it is changing it must, he says, be either in state *AB* or in state *BC* or partly in one and partly in the other. But neither the first nor the second alternative can apply while the thing is changing, and the third cannot apply either if the thing has no parts. Hence a thing with no parts cannot move, or change in any other way (240^b20–31). So far this appears to be just the argument that we have already criticized, though it may be noted that the symbolism '*AB*' and '*BC*' is evidently intended to suggest states that are in some way 'adjacent'.

A little more light is shed on the issue by Aristotle's subsequent remarks. A thing without parts could move, he says, only if time were made up of instants; for then there would be no time when it *is* moving, but only times when it *has* moved. But this has already been shown to be impossible (240^b31–241^a6). Moreover, he adds, before anything can move a distance greater than itself it must first move a distance less than or equal to itself. But such a thing as a point, being indivisible, cannot move a distance less than itself; so, if it can move at all, it must be able to move a distance equal to itself, and that in turn implies that a line may be made up of points; for the one point, by moving first one and then another distance equal to itself, will eventually measure out the whole line. But we have already shown that that is impossible (241^a6–14). Elaborating these thoughts further, Aristotle adds that it must in fact always be possible for anything that moves to move a distance less than itself, so clearly a point cannot move. What is indivisible and without parts, he concludes, could move only if it were possible to move in an indivisible instant, 'for movement in an instant, and the movement of what is indivisible, depend on the same reasoning' (241^a15–26).

One can see from this what position Aristotle considers himself to be controverting. When he considers the movement of an 'indivisible thing', or a 'thing without parts', he nevertheless conceives that thing as still having some size. That is why he naturally uses the symbol '*AB*', which represents an *interval*, to stand in for its initial position, and '*BC*', which represents an adjacent interval of the same length, to stand in for its 'next' position. It is also why he simply takes it for granted that we can meaningfully speak of a distance greater than, equal to, or less than the thing itself, i.e. a stretch comparable to the stretch it occupies itself. Although he calls such a thing a point, it is perfectly clear that his argument does not in fact show that a genuine point, which occupies *no* spatial distance, is incapable of moving. At best it shows that if an extended thing can move (in the direction in which it is extended), and if motion is continuous rather than 'cinematographic', then that thing must be divisible, at least in thought. But if we consider a genuinely unextended thing, such as the point at the tip of an arrow, it is perfectly evident that this can move, even though it has no parts. (Similarly, a surface of a cube can move in a direction in which it is not extended.)

Now Aristotle no doubt thinks that he has a reply to this objection, in that he has already conceded that a thing without parts may move *per accidens*, in so far as it belongs to some whole which moves *per se*. But first, this concession in no way improves his argument, for that argument did not anywhere rely on the premiss that only motion *per se* is in question. (Hence, if the argument were successful against motion *per se*, it would also be successful against motion *per accidens*, which clearly it cannot be.) Second, the point that he has in mind seems in fact to be a point about the identification of things without parts, and not about their motion. No doubt it is plausible to say that we can only identify the point of the arrow as the *same* thing, now in one place and now in another, if we identify it by reference to the whole arrow (or at least, some extended part of the arrow) of which it is a point. (This, one might suggest, is because we identify things in the first place by what we can perceive, and unextended things do not, 'in their *own* right', have any perceptible properties.[22]) But then the correct conclusion to draw seems to be that an unextended thing cannot be identified *per se*, though there would be no difficulty in supposing it to move *per se*. For there is not, as Aristotle apparently claims, anything in the concept of motion that requires a moving thing to have parts. (Thus if, for example, the material world were wholly made up of unextended centres of force—as Boscovich suggested—there would be no conceptual impediment to their moving.)

6. CONCLUDING REMARKS

We have seen that quite a number of the claims that Aristotle makes in book VI of the *Physics* are mistaken. Is there anything that can be said in explanation of these mistakes? No doubt it is sometimes right just to say that he was being careless, as with the error with which the book opens, that he fails even to state the definition of a continuum on which his first argument relies. But this seems an implausible view about most of the theses that we have disputed, for generally Aristotle offers quite extended arguments in their favour.

One suggestion might be that for much of the time he is responding, somewhat over-hastily, to issues raised by Plato in the second part of his *Parmenides*.[23] For example, the argument purporting to show that an indivisible thing cannot move is taken straight from the *Parmenides* (at 138d1–e7), though of course this does not make it any better. More significantly, Plato's main contribution in the *Parmenides* was certainly to point to the puzzling nature of an instantaneous change, and perhaps Aristotle's claim that no change is instantaneous is mainly an attempt to avoid the puzzles that Plato had revealed. Some support for this conjecture may perhaps be drawn from an earlier passage in

[22] But is 'being now' a perceptible property of an instant? And can we identify 'one and the same now', as it 'moves' through time, without first identifying anything else? (Contrast 219^b15–33.)

[23] I have discussed Plato's contribution in my (1978).

chapter 2 of book V, where again it is very tempting to see the influence of the *Parmenides*.

Plato had suggested that if we consider a change say from rest to motion, or from motion to rest, we can only say *when* this change occurs by positing a paradoxical 'sudden', not itself in time at all, for it to occur in. For only 'in a sudden' can a thing be neither in motion nor at rest (156c1–e7). He had then gone on to generalize this result to all other changes, and had included here not only the change from being *F* to being not-*F* but also the changes from being *F* to *becoming* not-*F*, and from there to being not-*F*. These too, he had suggested, must occur 'in a sudden', for only 'in a sudden' can a thing be neither (still) *F* nor (already) *becoming* not-*F*, and neither (still) becoming not-*F* nor yet (already) not-*F* (156e7–157b3). He thus raises a question not only about the instantaneous change from being *F* to being not-*F*, but also about the equally instantaneous change from being *F* to *changing* from being *F* to being not-*F*, and clearly there is the potential here for an infinite regress. Aristotle's response, then, is to deny in chapter 6 of book VI that the change from being *F* to being not-*F* is instantaneous, and to deny in chapter 2 of book V that there is such a thing as the change from being *F* to being in the process of becoming not-*F*. (For clearly if there were such a change it would have to be instantaneous.) In the course of this latter response (225^b13–226^a23) he shows himself well aware of the infinite regress that would be involved (225^b33–226^a6).

(Unfortunately, to support his position in chapter 2 of book V Aristotle falls back upon the narrow ontology of the *Categories*, according to which changes are not substances, but only substances can be the subjects of accidental predications, and hence only substances can undergo change. This leads him to generalize his response to the somewhat astounding claim that there is *no* kind of change of change—for example, a change cannot increase or decrease—and so to deny any subject matter to the study of acceleration and deceleration (225^b13–21). But the mistaken claims of book VI do not seem to rely on any such narrow ontology; for example, there would appear to be nothing in book VI that would conflict with the doctrine of book IV that change is indeed the subject of which time is predicated (contrary to *Categories* 2^a34–$^b6$).[24] At most there is a hint of the preference for substances over other things when Aristotle suggests—without any evident warrant—that it is the divisibility of the changing object that is the foundation for the divisibility of space and time (235^b1–3).)

But if it is Plato's *Parmenides* that provides the motivation for Aristotle's claim that every change takes time, one would certainly expect him to have worked out the consequences of this claim rather more carefully. In particular, one would expect him to have explained in more detail how change between contradictories

[24] Yet book IV would appear to claim that an instantaneous change is the subject of which the now is predicated, as I have argued in my (1980), Essay 9 in this volume. If so, it is of course in conflict with book VI, which denies the existence of instantaneous changes.

could be harmonized with his thesis, for at first sight it certainly is a counter-example, and indeed the counter-example from which Plato's own argument begins. And although we have in fact been able to suggest a possible reconstruction of Aristotle's thought upon this issue, we have had to read some way between his lines to do so, which makes it altogether less probable that his account is meant as an explicit reply to Plato. It appears, rather, that he simply has not grasped the problem posed by change between contradictories. (And we may note that when elsewhere he rejects the thesis that all change takes time, the counter-example that he has in mind seems not to be just any old change between contradictories, but the very special case of a change from not existing to existing, or vice versa, in the case of things that have no parts.)

Somewhat similarly, one might suggest that Aristotle's claim that there is no sense to talk of motion or rest at an instant is intended as a response to Plato's puzzle, for that puzzle begins by asking how there can be an instant at which a thing is neither in motion nor at rest, but is changing between them. But Plato generalizes his puzzle to all other changes, mentioning amongst others change of size, and change between existence and non-existence (156e7–157b3). So we may similarly ask how there can be an instant at which a thing neither exists nor does not exist, neither is of this size nor is not of this size. Aristotle, it would appear, should again be replying that there is no sense to talk of existence or non-existence at an instant, or of being a particular size just at one instant, but evidently he says no such thing. He denies that a thing either is changing at an instant or is not changing at it (counting moving as a kind of changing), but he makes no attempt to extend this to other apparent contradictories between which change may occur. So again it seems probable that this claim of his is not a response to Plato's puzzle, for if it were it would not be nearly general enough.[25]

Another possible suggestion, to explain Aristotle's denial of motion or rest at an instant, might be that he thought it was needed to resolve Zeno's paradox of the flying arrow. Certainly, chapter 8 concentrates upon the implications of the claim that nothing rests at an instant, and this then leads straight into a refutation of the arrow paradox at the beginning of chapter 9. But first, it would seem to be a somewhat exaggerated response to that paradox, and second, it appears to be only part of Aristotle's own response. Zeno apparently had argued that an arrow in its flight is always 'in some instant', and that in each instant it is 'over against what is equal to itself', from which he had inferred that in each instant it is at rest (239^b5–7). Now all that is needed by way of response to this is to point out that the inference is invalid. One can agree that at each instant a thing is always 'over against what is equal to itself' (which I take to be a way of saying that it occupies exactly the position that it does occupy), while denying that it follows from this that it is at rest (as Aristotle has just done, at 239^a35–$^b1$).

[25] One might indeed wonder whether my presumed chronology is correct. Perhaps Plato's *Parmenides* replies to Aristotle's *Physics* VI, rather than vice versa?

And surely it would not be too difficult to explain this denial while still leaving open the possibility of saying that at each instant of its flight the arrow is in motion. One is not forced to say that nothing at all counts as being in motion or at rest at an instant, which is what Aristotle gives as his reason at 239^b1–2; and it is at first somewhat difficult to see why he should add, as a further diagnosis, that it is false to suppose that time is made up of indivisible instants (239^b8–9). What has that to do with it?

It may be suggested that by this last remark Aristotle means to indicate that in any case the inference from 'the arrow rests at each instant' to 'the arrow rests throughout the period' is invalid. His point is that it is invalid *because* time is not made up of instants. (For even though Aristotle has already rejected the premiss, he still needs to reject the inference too. At any rate, he himself accepts 'the arrow covers no distance in any instant', but will not infer 'the arrow covers no distance throughout the period'.) The trouble with this suggestion is, of course, that the inference would still be invalid even if time *were* made up of instants, as we now believe that it is, and as the time-atomist also believes, but in a different way. But this leads to a better suggestion, for if we construe an 'instant' as a very short *period*, as the time-atomist does, then both the premiss and the inference may indeed seem tempting. At any rate, both we and Aristotle would accept as correct 'if the arrow covers no distance in any millisecond of this minute, then it covers no distance during the whole minute'. Similarly, both we and Aristotle would agree that if the arrow were at the same spatial position throughout a whole millisecond then it could properly be said to be at rest then. I suggest, then, that Aristotle thought that the plausibility of this paradox depends upon the time-atomist's way of construing an instant as a short period, and that is why his diagnosis is that it is a mistake to suppose that time is made up of instants. He means that time-atomism is a mistake, and he takes this thesis to *be* a characterization of time-atomism.

Now in fact the consistent time-atomist should accept that the arrow covers no distance in any time-atom, though he is not bound to conclude that it therefore counts as being 'at rest' then. But in any case he should certainly reject the suggested inference, since it depends on the assumption that motion is not 'cinematographic'. Thus the paradox appears to have some bite only if one is inconsistently combining some features of the theory that time is atomic with some features of Aristotle's own theory that time is a continuum. But the significant point that emerges is that, from Aristotle's point of view,[26] It is time-atomism that lies behind this paradox. And I would suggest that it is not this paradox in particular but time-atomism in general that leads him to deny instantaneous motion or rest. In fact it is plausible to say that atomism, in one form or another, is Aristotle's main target in this book of the *Physics*. At any rate,

[26] Whether this was also Zeno's view is another question.

it is certainly his target in all of chapters 1–4, and this consideration will be enough to yield a fairly plausible explanation of all our difficulties.

We may start by repeating the point just made, that Aristotle regards the thesis that time is made up of instants as characteristic of time-atomism. Similarly, when the book opens with a refutation of the claim that a line is made up of points, Aristotle regards this claim as characteristic of space-atomism. For who, in his day, had claimed that lines *were* made up of points? Not, presumably, some unknown Cantor or Dedekind, with a theory about how it really was possible to construct a line from genuinely unextended points, and who had shown how, if you take *enough* things of no magnitude, you thereby obtain a positive magnitude. No. The thesis can only have seemed attractive to those who held—e.g. for reasons such as Aristotle gives at *De Generatione et Corruptione* 315^b24–316^b34—that points must after all be conceived as having magnitude, and yet as being indivisible. We would more naturally say that such things should not be called points at all (but, say, 'atoms of extension'), and Aristotle would no doubt agree. Yet as we have seen he is prepared to call them points himself, in the course of elaborating a different argument against them at the end of chapter 10 (241^a6–26). It is this theory, then, that he takes himself to be refuting at the beginning of chapter 1. His argument is, in effect, that if these things really are indivisible, then they can only be points as we and he conceive points to be. And it is hardly surprising that he cannot see how *genuine* points, which cannot touch one another, could be put together, to form a gapless continuum.

The rest of chapter 1 is then concerned to maintain the continuity of (uniform, rectilinear) motion, and on that ground to deny the atomicity of space and of time. This argument continues also throughout chapter 2. It is a very nice argument, given the premiss about the continuity of motion, but that premiss itself leads to a problem. For what argument can Aristotle use against the rival 'cinematographic' view of motion? The truth is that there is no good argument, and the only argument that he can come up with is this: we ordinarily suppose that if a thing *has* moved to a certain place, then at some earlier time it *was* moving to that place, whereas it can fairly be said that the 'cinematographic' view of motion denies this. So Aristotle has to elevate this 'ordinary supposition' of ours to an undeniable principle. But now, would there be any way of explaining, without begging the question, why this principle should hold of motion from place to place but should *not* hold of other changes? Certainly, it is difficult to think of one, and it would appear that we do ordinarily make this same supposition whatever kind of change is in question. Thus Aristotle is apparently committed to saying that in *all* cases 'has changed' implies 'was changing', and this of course is exactly what he does say in chapter 6. It is, then, his way of arguing against the atomicity of space and time in chapter 1 that commits him to his claim in chapters 5 and 6 that *all* change takes time. (He may of course also have been influenced by the view which he argues later in chapter 7 of book VIII, that all other kinds of change depend upon motion from place to place.) But I think one can only conclude that he has not really seen the difficulties that this claim involves.[27]

Anyway, to return to our thread, chapters 1 and 2 clearly argue against 'atomist' theories of space and time, and it is evidently a continuation of the same theme when Aristotle proceeds, in chapter 3, to give further reasons for saying that 'a now' has no temporal extension. It is still, I think, part of the same programme when he proceeds to infer from this that movement and rest cannot occur 'in an instant'. For, as he conceives it, the idea that they can is an atomist idea; it is part of the idea that movement proceeds 'cinematographically'. That is why, when he uses this thesis in his reply to Zeno's arrow, he presumes that Zeno must be basing his argument on an atomist premiss. This conjecture would also go some way towards explaining an otherwise puzzling passage in chapter 5, where Aristotle appears at first sight to be rejecting his own claim that nothing can be said to be at rest at an instant.

At this point Aristotle is opening his argument for the claim that there is no first time in which a thing has changed, that is, has begun its change, or has changed at least a bit. For suppose, he says, *AD* is such a time. Then *AD* will be divisible into an earlier and a later part, and clearly the earlier part will equally be a time in which the thing has begun its change, or has changed at least a bit. But before he draws his conclusion, he pauses to consider the rival claim that *AD* is not divisible, and is therefore a time-atom. As one objection to this, he says that it will follow that nows (i.e. time-atoms) will be next to one another, whereas he has already argued that nows (genuine instants) cannot be (236^{a}15–17). But he also offers a second objection, which is puzzling. We may assume that the thing was at rest throughout a preceding period *CA*. But then, he goes on, 'it was at rest at *A*; so if *AD* is without parts it will simultaneously be at rest and will have changed, for it was at rest at *A*, but has changed at *D*' (236^{a}17–20). What is puzzling about this is that Aristotle appears to be assuming, as a *premiss* to his objection, that one can properly speak of the thing as being at rest at the instant *A*, although this premiss denies his own doctrine. So the explanation can only be (as suggested by Sorabji[28]) that he is not asserting this premiss in his own person, but is taking it as one that his opponent, the time-atomist, would have to grant. But why so? Well, perhaps the answer is this. The time-atomist supposes that during a now (i.e. a time-atom) everything remains unchanged. So in order to express his thesis at all, *he* must suppose that it makes sense to say that a thing is at rest in a now (i.e. a time-atom), and Aristotle—having 'proved' that a now is not extended—takes this to imply that the opponent is committed to saying that a thing may be at rest in a genuine instant. Thus his argument is that the time-atomist could be forced to accept that we can meaningfully ask whether a thing is or is not at rest *at* the instant which separates one supposed time-atom from the

[27] And hence that he was *not* familiar with *Parmenides* 156–7?

[28] See n. 16 above.

next, and he has already claimed that this question cannot be answered when the instant in question is one that separates a period of rest from a period of motion (234^{a}34–^{b}5).

Needless to say, this is not a good argument against time-atomism, for the time-atomist may perfectly well refuse to say what state anything is in *at* the instant that separates one time-atom from the next, and anyway the argument confuses the indivisibility that applies to a time-atom (which does not exclude divisibility 'in thought') with the indivisibility of a genuine instant. But, if this reconstruction is right, then we may note two ways in which time-atomism may be taken to require motion and rest in an instant. It requires that a thing may *have moved*, in a genuine instant, from place *A* to place *B*, for that is how cinematographic motion occurs. It also requires that things *have remained at rest* during each time-atom, for that is precisely what time-atomism claims, and Aristotle takes this to imply that a thing has remained at rest in a genuine instant. These, then, are the theses that Aristotle really wishes to deny.

But if this is right, then we can only say that he has been taken in by an ambiguity, an ambiguity between 'nothing moves in an instant' and 'nothing is moving at an instant', for indeed the Greek for both is exactly the same: *οὐθὲν κινεῖται ἐν τῷ νῦν* (Greek does not distinguish the so-called continuous present tense 'is moving' from the plain present tense 'moves'; and it standardly uses 'in an instant' where we would say 'at an instant'.) Undoubtedly, Aristotle's first argument for this claim is only an argument for the first version of it, for it professes to deduce a contradiction from the supposition that a thing *has* moved from *A* to *B* during one indivisible instant (234^{a}26–31). And there is no recognition that the sense of the claim has shifted in the following argument (234^{a}34–^{b}5), where however it is much more natural to suppose that what is in question is whether a thing can be said to *be* moving *at* an instant. At any rate, it is clear that he does employ the claim in this latter sense (e.g. at 237^{a}11–15 and 239^{a}10–14), though he also thinks, as we have seen, that only an atomist would wish to deny it. So there is some confusion in his thought here, and the only version of the claim that he actually needs is the less controversial one: nothing can be said either to move from *A* to *B*, or to remain at *A*, in an instant.

Finally, chapter 4, with its claim that a moving *thing* must always be divisible, is all part and parcel of this same attack on atomism, though the attack has moved now from atomism about space and time to the more familiar atomism about matter. (Though this may not be obvious from the way that Aristotle presents his reasoning in chapter 4, it is obvious from his further elaboration in chapter 10.) He wishes to claim that an atomic chunk of matter, if it really is (as its proponents say) not divisible into parts, cannot move at all. In one way he is right: if an atom is an extended chunk of matter, and if motion is continuous and not cinematographic, then it must be possible for a moving atom to be partly in and partly not in the place it occupied a little earlier. And in this sense it must have parts. But (i) it was a mistake on Aristotle's part to let this claim depend upon the hypothesis that the atom moves, for an extended atom must still have spatial parts even if it never moves, and there is nothing in the concept of motion that requires moving things to be extended; and (ii) Aristotle's objection anyway misses the point. Just as an atomist concerning space or time will conceive of his spatial or temporal atoms as extended, so that they can always be regarded as divisible *in thought*, so also one who believes in atoms of matter will equally regard them as extended, and should therefore admit that they are equally divisible *in thought*. His claim is rather that they are never divided in fact, that nothing that you can do to an atom will ever separate it into parts that move independently of one another. But against this theory Aristotle never offers any objection that carries any conviction at all.

To sum up the situation a little crudely, Aristotle has three kinds of objection to atomism about matter. One is that the Greek atomists thought of atoms as differing from one another only in size and shape, and so attempted to explain all the various properties of material objects by drawing just on these properties of their constituent atoms. Aristotle very fairly objects that such a reduction cannot be made to work (e.g. at *De Caelo* III. 4, $303^{a}3$–$^{b}3$, and III. 8, $306^{b}29$–$307^{b}24$). But clearly this is not an objection of principle, since it could apparently be met by endowing atoms with further properties. His second objection is that the atomist hypothesis must also involve positing the existence, of empty space, and this he thinks he can refute independently (*Physics* IV. 6–9). But his third objection, which he appears to think of as the strongest, is what he calls the conflict with mathematics, and it is this objection that is developed in *Physics*. VI. (Cf. *De Caelo* III. 4, $303^{a}20$–4 and III. 7, $306^{a}26$–$^{b}2$.) But, as we have seen, it misses the point altogether. One can be an atomist concerning space, time, or matter, and still admit the *mathematical* divisibility of one's atoms.

Additional Note 10.1

There is nothing that I wish to subtract from this essay, save perhaps to admit that I now think that it does 'protest too much' on the way to its conclusion that in *Physics* VI Aristotle's main aim is to argue against all forms of atomism, i.e. atomisms of space, time, motion, and matter. For, so far as I am aware, this conclusion is not seriously controversial. One question which the essay raises, but only in footnotes (i.e. notes 23, 25, 27), is the relation between this book of Aristotle's *Physics* and the treatment of closely related issues in the second part of Plato's *Parmenides*. I am still inclined to think that it is not at all obvious which of these was written first and which second, or whether either of them should be regarded as responding to the other. But that is not a question that I can here pursue further.

Instead I merely add a further comment on Aristotle's attempt to reject all forms of atomism, but now focused on an argument that he discusses, not in the

Physics, but in *De Generatione et Corruptione* I. 2, at 316^a15–317^a17. This argument may fairly be called 'the paradox of division everywhere', and it goes roughly like this. On Aristotle's view a continuum is 'divisible everywhere', from which it appears to follow that it must be possible that it should be everywhere divided. But then we cannot say what would be left when this division was completed. The conclusion that we are invited to draw is that division everywhere cannot be possible, so there must be some indivisible stretches, i.e. extended atoms (whether of space, time, or matter, for the argument applies equally well to each case). I shall not be concerned here with the detail of Aristotle's own discussion, since that is quite often obscure, and I think that there is no one who would wish to defend it. But the problem is genuine enough.[29]

The Paradox of Division Everywhere

It is important to begin by noting that there are actually two quite different problems that go under this title. The first does indeed concern division *everywhere*, and goes like this. A (finite) line is supposed to be divisible at every point. Very well then, suppose it is actually divided at every point, and ask what we have left. They cannot be segments with magnitude, for if they are the division will not have been completed; but equally they cannot be parts without magnitude, for no number of parts without magnitude could compose a whole which has magnitude. The second version (which is apparently Zeno's version) does not literally concern division *everywhere*, but a specially selected infinity of divisions. It asks us to imagine a line divided into halves, and then each of those halves divided into further halves, and so on ad infinitum. This introduces a countable infinity of divisions, but not a division *everywhere*. However, the problem posed seems to be just the same. For any segment of the line which does have magnitude will have been divided by one of the specified divisions.

In the light of Aristotle's claim in *Physics* III that all infinity is merely potential and never actual, one might expect him to reply that this apparent paradox arises only because the opening premiss seduces us into trying to make sense of a situation which really is impossible. For, in either version of the paradox, 'division everywhere' would clearly require an infinity of divisions, and Aristotle holds that no such infinity can ever be completed. The idea would be that the divisions must be thought of as happening one after another, in an unending series. At any stage within this series we can quite coherently describe the situation then reached, and can see that further divisions are still possible. But since it is impossible ever to reach a position in which *all* the stages have been completed, it is not surprising that we cannot say what the situation would then be. It is interesting that this is very clearly *not* the response that Aristotle himself offers,

[29] What follows is mainly excerpted from my (1988).

when he discusses this argument in *De Generatione et Corruptione* I. 2.[30] This is in a way a good thing, for as I have argued in Essay 7 there are infinite series which can be completed, so if this one cannot some further explanation of why not is needed. But unfortunately one cannot make any good sense out of what Aristotle does say by way of explanation (which is apparently based on the surely irrelevant observation that no two points can be next to one another), so I henceforth consider the problem with no further reference to him.

With the first version of the puzzle, in which we are supposed to be dividing *everywhere*, it will make a difference how we are to conceive of a division. Suppose, for example, that dividing a line at a point is thought of as simply taking away that point. Then obviously to divide the line at every point is just to take away every point, and so leaves nothing at all remaining. But suppose instead that a division is thought of rather as inserting a gap, separating the point in question from (say) its left-hand neighbours. The gap is to be a (half-open) interval of some finite length, and as a result of the division the distance between the end-points of the original line is increased by that length. Then the answer is that we cannot divide everywhere, but the reason is quite sophisticated, and goes beyond anything that Aristotle might have known. It is that in order to divide by inserting gaps at *every* point we should have to insert an uncountable infinity of gaps, and the original end-points would then be separated by an uncountable infinity of non-overlapping line segments. But (assuming the space to be Archimedean) that is impossible. There is in fact no way of spreading out the points of a finite line so that each one of them becomes separated by some definite gap from all the others.

The second version of the puzzle avoids this problem, since the second version only calls for a countable infinity of divisions. Thinking of a division in the first way, we simply remove a countable infinity of points, so evidently we leave plenty of points remaining. But the points we have removed are dense in the line, which means that any two remaining points are separated by a removed point, and hence no complete line-segment of any length remains. Yet the remaining points, taken together, still have the same *measure* as the original line. The case is entirely similar if we divide by inserting gaps. To make this possible, we must of course suppose that the gaps inserted are successively smaller and smaller, for if each gap had the same size then any two points of the original line would end up an infinite distance away from each other, and that is impossible. But there is no difficulty if the gaps diminish appropriately, and we may, for example, arrange things so that the whole operation merely doubles the original distance between the end-points. Then, as before, there will be no complete line-segment remaining, but we will not have separated any point of the original line by any definite gap from all of its former neighbours. That is, for any point of the line,

[30] I take this to be a strong ground for saying that, just as *Physics* VI must have been written before the account of infinity in *Physics* III, so too this discussion in *GC* I. 2 must have been earlier.

and any distance you care to take, however small, there will still be points of the original line within that distance of the given point.

Nevertheless, with either version of the division, it does seem reasonable to conclude that we have 'broken up' the original line into 'nothing but points', and points have no magnitude. To resolve the paradox, therefore, it is necessary to deny Aristotle's claim that a line cannot be made up out of points, and to explain how lines with magnitude can indeed be put together from points without magnitude, provided there are 'enough' of them. 'Enough' here must of course be more than any finite number, but also it must be *more* than a countably infinite number. So it is perhaps hardly surprising that the Greeks never did reach an adequate solution to this puzzle, and for them it remained a strong argument for atomism.

References

ACKRILL, J. L. (1972–3), 'Aristotle's Definitions of *Psuchē*', *Proceedings of the Aristotelian Society*, 73: 119–33; repr. in Barnes *et al.* (1979).

——(1981), *Aristotle the Philosopher* (Oxford: Oxford University Press).

ANNAS, J. (1975), 'Aristotle, Number, and Time', *Philosophical Quarterly*, 25: 97–113.

ANSCOMBE, G. E. M., and GEACH, P. T. (1963), *Three Philosophers* (Oxford: Blackwell).

BALME, D. J. (1962), 'Development of Biology in Aristotle and Theophrastus: Theory of Spontaneous Generation', *Phronesis*, 7: 91–104.

——(1972), *Aristotle*: De Partibus Animalium *I and* De Generatione Animalium *I* (Oxford: Clarendon Press).

——(1975), 'Aristotle's Use of Differentiae in Zoology', in Barnes *et al.* (1975).

——(1980), 'Aristotle's Biology Was Not Essentialist', *Archiv für Geschichte der Philosophie*, 62: 1–12.

——(1987), 'Teleology and Necessity', in Gotthelf and Lennox (1987).

BARNES, J., SCHOFIELD, M., and SORABJI, R. (eds.) (1975), *Articles on Aristotle*, i: *Science* (London: Duckworth).

————————(1979), *Articles on Aristotle*, iv: *Psychology and Ethics* (London: Duckworth).

BONITZ, H. (1870), *Index Aristotelicus* (Berlin).

BOSTOCK, D. (1978), 'Plato on Change and Time in the *Parmenides*', *Phronesis*, 23: 229–42.

——(1979), *Logic and Arithmetic*, ii (Oxford: Clarendon Press).

——(1980), 'Aristotle's Account of Time', *Phronesis*, 25: 148–69. (Essay 9 in this volume.)

——(1982), 'Aristotle on the Principles of Change in *Physics* I', in Schofield and Nussbaum (1982). (Essay 1 in this volume.)

——(1988), 'Time and the Continuum', *Oxford Studies in Ancient Philosophy*, 6: 255–70.

——(1991), 'Aristotle on Continuity in *Physics* VI', in Judson (1991b). (Essay 10 in this volume.)

——(1994), *Aristotle*: Metaphysics, *Books Z and H* (Oxford: Clarendon Press).

——(1995), 'Aristotle on the Transmutation of the Elements in *GC* I: 1–4', *Oxford Studies in Ancient Philosophy*, 13: 217–29. (Essay 2 in this volume.)

——(1996), See Waterfield (1996).

——(2004), 'An Aristotelian Theory of Predication?', *Oxford Studies in Ancient Philosophy*, 27: 141–75.

BOYLAN, M. (1981), 'Mechanism and Teleology in Aristotle's Biology', *Apeiron*, 15: 96–102.

BRADIE, M., and MILLER, F. D. (1984), 'Teleology and Natural History in Aristotle', *History of Philosophy Quarterly*, 1: 133–46.

CALLAHAN, J. F. (1948), *Four Views of Time in Ancient Philosophy* (Cambridge, Mass.: Harvard University Press).

CHARLES, D. (1988), 'Aristotle on Hypothetical Necessity and Irreducibility', *Pacific Philosophical Quarterly*, 69: 1–53.
—— (1991), 'Teleological Causation in the *Physics*', in Judson (1991*b*).
CHARLTON, W. (1970), *Aristotle's* Physics, *Books I and II* (Oxford: Clarendon Press).
COBURN, R. (1971), 'Identity and Spatio-temporal Continuity', in M. Munitz (ed.), *Identity and Individuation* (New York: New York University Press).
CODE, A. (1976), 'The Persistence of Aristotelian Matter', *Philosophical Studies*, 29: 357–67.
COHEN, S. M. (1984), 'Aristotle's Doctrine of the Material Substrate', *Philosophical Review*, 93: 171–94.
—— (1996), *Aristotle on Nature and Incomplete Substances* (Cambridge: Cambridge University Press).
COOPER, J. M. (1982), 'Aristotle on Natural Teleology', in Schofield and Nussbaum (1982).
—— (1985), 'Hypothetical Necessity', in Gotthelf (1985*b*).
—— (1987), 'Hypothetical Necessity and Natural Teleology', in Gotthelf and Lennox (1987).
DIELS, H., rev. KRANZ, W., *Die Fragmente der Vorsokratiker*, 10th edn. (Berlin: Weidmannsche Verlagsbuchhandlung, 1961).
DÜRING, I. (1944), *Commentary on Aristotle's* Meteorologica *IV* (first pub. in *Göteborgs Högskalas Årsskrift*, 50; repr. New York: Garland, 1980).
FREDE, M. (1978), 'Individuals in Aristotle', tr. from German in Frede (1987).
—— (1985), 'Substance in Aristotle's *Metaphysics*', in Gotthelf (1985*b*); repr. in Frede (1987).
—— (1987), *Essays in Ancient Philosophy* (Oxford: Oxford University Press).
FREUDENTHAL, G. (1995), *Aristotle's Theory of Material Substance: Heat and Pneuma, Form and Soul* (Oxford: Oxford University Press).
FURLEY, D. J. (1983), 'The Mechanics of *Meteorologica* IV', in P. Moraux and J. Wiesner (eds.), *Zweifelhaftes im Corpus Aristotelicum* (Berlin: De Gruyter).
—— (1985), 'The Rainfall Example in *Physics* II. 8', in Gotthelf (1985*b*).
—— (1996), 'What Kind of Cause is Aristotle's Final Cause?', in M. Frede and G. Striker (eds.), *Rationality in Greek Thought* (Oxford: Oxford University Press).
FURTH, M. (1988), *Substance, Form and Psyche: An Aristotelian Metaphysics* (Cambridge: Cambridge University Press).
GEACH, P. T. (1962), *Reference and Generality* (Ithaca, NY: Cornell University Press).
GERSHENSON, D. E., and GREENBERG, D. A. (1961), 'Melissos of Samos in a New Light: Aristotle's *Physics* 186^a 10–16', *Phronesis*, 6: 1–9.
GILL, M. L. (1989), *Aristotle on Substance* (Princeton: Princeton University Press).
GOTTHELF, A. (1976), 'Aristotle's Conception of Final Causality', *Review of Metaphysics*, 30: 226–54; repr. with additions in Gotthelf and Lennox (1987).
—— (1980), Review of Nussbaum (1978), *Journal of Philosophy*, 77: 365–78.
—— (1985*a*), 'Notes towards a Study of Substance and Essence in Aristotle's *Parts of Animals* II–IV', in Gotthelf (1985*b*).
—— (ed.) (1985*b*), *Aristotle on Nature and Living Things* (Bristol: Bristol Classical Press).

—— and LENNOX, J. G. (eds.) (1987), *Philosophical Issues in Aristotle's Biology* (Cambridge: Cambridge University Press).

GRAHAM, D. W. (1987), 'The Paradox of Prime Matter', *Journal of the History of Philosophy*, 25: 475–90.

HARDIE, R. P., and GAYE, R. K. (1930), *Aristotle's* Physics, in W. D. Ross (ed.), *The Works of Aristotle Translated into English*, ii (Oxford: Clarendon Press).

HIRSCH, E. (1982), *The Concept of Identity* (Oxford: Oxford University Press).

HUME, D. (1739), *A Treatise of Human Nature*, ed. L. A. Selby-Bigge, rev. P. H. Nidditch (Oxford: Clarendon Press, 1978).

IRWIN, T. (1988), *Aristotle's First Principles* (Oxford: Oxford University Press).

JONES, B. (1974), 'Aristotle's Introduction of Matter', *Philosophical Review*, 83: 474–500.

JUDSON, L. (1991*a*), 'Chance and "Always or for the Most Part" in Aristotle', in Judson (1991*b*).

—— (ed.) (1991*b*), *Aristotle's* Physics*: A Collection of Essays* (Oxford: Oxford University Press).

KING, H. R. (1956), 'Aristotle without *Materia Prima*', *Journal of the History of Ideas*, 17: 370–87.

LEAR, J. (1988), *Aristotle: The Desire to Understand* (Cambridge: Cambridge University Press).

LENNOX, J. G. (1982), 'Teleology, Chance, and Aristotle's Theory of Spontaneous Generation', *Journal of the History of Philosophy*, 20: 219–38.

—— (2001*a*), *Aristotle:* On the Parts of Animals *I–IV* (Oxford: Clarendon Press).

—— (2001*b*), *Aristotle's Philosophy of Biology* (Cambridge: Cambridge University Press).

LEWIS, F. A. (1988), 'Teleology and Material/Efficient Causes in Aristotle', *Pacific Philosophical Quarterly*, 69: 54–98.

—— (1991), *Substance and Predication in Aristotle* (Cambridge: Cambridge University Press).

MANSION, A. (1945), *Introduction à la physique aristotélicienne*, 2nd edn. (Louvain: Édition de L'Institut supérieur de philosophie).

MATTHEWS, G. B. (1982), 'Accidental Unities', in Schofield and Nussbaum (1982).

MILLER, F. D. (1974), 'Aristotle on the Reality of Time', *Archiv für Geschichte der Philosophie*, 56: 132–55.

MORAVCSIK, J. M. E. (ed.) (1967), *Aristotle: A Collection of Critical Essays* (New York: Anchor Books).

MORISON, B. (2002), *On Location* (Oxford: Oxford University Press).

NEWTON, I. (1686), *Philosophiae Naturalis Principia Mathematica*, tr. A. Motte, rev. F. Cajori (Berkeley and Los Angeles: University of California Press, 1962).

—— (1704), *Opticks* (NewYork: Dover, 1952).

NUSSBAUM, M. C. (1978), *Aristotle's De Motu Animalium* (Princeton: Princeton University Press).

OWEN, G. E. L. (1967), 'Tithenai ta Phainomena', in Moravcsik (1967).

—— (1976), 'Aristotle on Time', in P. Machamer and R. Turnbull (eds.), *Motion and Time, Space and Matter* (Columbus: Ohio State University Press).

PECK, A. L. (1942), *Aristotle:* Generation of Animals (Cambridge, Mass.: Harvard University Press).

PECK, A. L. (1945, 1961), *Aristotle:* Parts of Animals (Cambridge, Mass.: Harvard University Press).
PELLEGRIN, P. (1985). 'Aristotle: A Zoology without Species', in Gotthelf (1985*b*).
ROBINSON, H. M. (1974), 'Prime Matter in Aristotle', *Phronesis*, 19: 168–88.
ROSS, W. D. (1936), *Aristotle's* Physics (Oxford: Clarendon Press).
—— (1955), *Aristotelis Fragmenta Selecta* (Oxford: Clarendon Press).
—— (1961), *Aristotle:* De Anima (Oxford: Clarendon Press).
SCHOFIELD, M., and NUSSBAUM, M. C. (eds.) (1982), *Language and Logos: Studies in Ancient Greek Philosophy Presented to G. E. L. Owen* (Cambridge: Cambridge University Press).
SEDLEY, D. (1991), 'Is Aristotle's Teleology Anthropocentric?', *Phronesis*, 36: 179–96.
SHOEMAKER, S. (1969), 'Time without Change', *Journal of Philosophy*, 66: 363–81.
SOLMSEN, F. (1958), 'Aristotle and Prime Matter', *Journal of the History of Ideas*, 19: 243–52.
SORABJI, R. (1976), 'Aristotle on the Instant of Change', *Proceedings of the Aristotelian Society, Supplementary Volume*, 50: 69–89.
—— (1980), *Necessity, Cause, and Blame* (London: Duckworth).
—— (1983), *Time, Creation and the Continuum* (London: Duckworth).
STAHL, D. (1981), 'Stripped Away', *Phronesis*, 26: 177–80.
STRAWSON, P. F. (1959), *Individuals* (London: Methuen).
SWINBURNE, R. (1981), *Space and Time*, 2nd edn. (London: Macmillan).
WARDY, R. (1993), 'Aristotelian Rainfall or the Lore of Averages', *Phronesis*, 38: 18–30.
WATERFIELD, R. (tr.) (1996), *Aristotle:* Physics, with introd. and notes by D. Bostock (Oxford: Oxford University Press).
WATERLOW, S. (1982), *Nature, Change, and Agency in Aristotle's* Physics (Oxford: Clarendon Press).
—— (1984), 'Aristotle's Now', *Philosophical Quarterly*, 34: 104–28.
WHITE, M. J. (1992), *The Continuous and the Discrete* (Oxford: Clarendon Press).
WIELAND, W. (1960–1), 'Aristotle's *Physics* and the Problem of Inquiry into Principles', tr. in Barnes *et al.* (1975).
—— (1975), 'The Problem of Teleology', in Barnes *et al.* (1975); trans. from his *Die Aristotelische Physik* (Göttingen: Vandenhoeck & Ruprecht, 1962).
WIGGINS, D. (1967), *Identity and Spatio-temporal Continuity* (Oxford: Blackwell).
—— (1980), *Sameness and Substance* (Oxford: Blackwell).
WILLIAMS, C. J. F. (1982), *Aristotle's* De Generatione et Corruptione (Oxford: Clarendon Press).
WITT, C. (1985), 'Form, Reproduction, and Inherited Characteristics in Aristotle's *GA*', *Phronesis*, 30: 46–57.
—— (1989), *Substance and Essence in Aristotle* (Ithaca, NY: Cornell University Press).
WOODS, M. J. (1982), *Aristotle's* Eudemian Ethics *I, II, VIII* (Oxford: Clarendon Press).

Index